Virtual Reality Technology in Mining Machinery

Xuewen Wang · Jiacheng Xie · Suhua Li

Virtual Reality Technology in Mining Machinery

Virtual Assembly, Virtual Planning and Virtual Monitoring

Springer

Xuewen Wang
College of Mechanical and Vehicle Engineering
Taiyuan University of Technology
Taiyuan, Shanxi, China

Jiacheng Xie
College of Mechanical and Vehicle Engineering
Taiyuan University of Technology
Taiyuan, Shanxi, China

Suhua Li
College of Mechanical and Vehicle Engineering
Taiyuan University of Technology
Taiyuan, Shanxi, China

ISBN 978-981-16-4407-8 ISBN 978-981-16-4408-5 (eBook)
https://doi.org/10.1007/978-981-16-4408-5

This Springer imprint is published by the registered company Springer Nature Singapore Pte Ltd.
The registered company address is: 152 Beach Road, #21-01/04 Gateway East, Singapore 189721, Singapore

Preface

Intelligent mining is the main content of promoting the high-quality development of coal industry. The innovative development of Internet +, big data, virtual reality and other advanced information technologies as well as the gradual industrial application of 5G and other advanced communication technologies have opened a new era of intelligence in the coal industry, which is currently in the research and development process of transparent working face.

Digital twin technology and other concepts are gradually integrated into the construction of Digital factories. In recent years, it has been gradually integrated with the fully mechanized mining production system. Based on sensors, big data and Internet of Things technologies, it is expected to realize the independent operation of the fully mechanized mining unit in the complex and changeable coal seam environment, and the operation process is completely transparent and visible. However, the operation of fully mechanized mining face has the following characteristics:

(1) The equipment group operates dynamically in real time under the condition of coal seam occurrence and the uneven environment of roof and floor;
(2) In the closed space, the movement relationship of equipment connection is complex. Except for coal shearer, it is difficult and costly to accurately position and set the attitude of all other equipment;
(3) It is difficult to accurately and real-time present the operating state of the production system on the working face, so as to achieve the effect of "what you see is real". The digital twin fully mechanized mining face is established, and the production system of the fully mechanized mining face is duplicated in the virtual environment, so that the virtual space model and physical space model of the production system of the fully mechanized mining face are in real-time interaction and can respond to the dynamic changes of each other in time, which lays the foundation for the realization of "unmanned" mining. The realization of all these cannot be achieved without virtual reality technology. It is one of the key technologies of digital twin technology to establish high-precision virtual scenes and establish virtual and real interaction channels in virtual environment. Therefore, the virtual assembly, virtual planning and virtual monitoring of coal machine equipment in the virtual environment is of

great significance to the establishment of digital twin fully mechanized mining face and the realization of intelligent mining.

The book consists of 14 chapters.

Chapter 1 is the application overview, which introduces the research background, purpose and significance of this book through the virtual reality assembly technology and system as well as the three machines in fully mechanized mining face monitoring and dynamic planning in the fully mechanized mining face under the virtual reality environment.

Chapter 2 is the overall design of the system, which proposes a design and operation mode of the production system of fully mechanized mining face based on digital twin. In fully mechanized design and monitoring service system, driven by factors of production management configuration, process simulation and real-time monitoring, etc. in the physical and virtual of fully mechanized working face, fully mechanized design and monitoring service system of fully mechanized working face of iterative operation, to achieve full elements, whole process, full integration and data fusion, to achieve the optimal configuration and equipment production system together; the purpose of safe and efficient mining. The technical framework and key technologies of the three stages are described in detail.

Chapter 3 introduces the virtual simulation element model conversion technology and the establishment of the model library, and in accordance with the internal structure and the working characteristics, optimize the technology of CAD model transformation and 3 on 3d Max scene and animation technology is established by virtual dismounting module and scene simulation module of virtual reality database, provides a foundation for the establishment of the system, and can continue to expand the relevant function later on down the line.

Chapter 4 puts forward the method of virtual assembly based on OSG, manipulation, recording and playback of the path, the model reset, automatic positioning constraints, network collaborative assembly and parts introduces the concrete methods of assembling and disassembling demo automatically carried out theoretical research and through the programming to realize the corresponding function, based on this method to establish the virtual assembly system, based on OSG can meet user demand for virtual assembly.

Chapter 5 puts forward the virtual assembly method based on UG and uses the product information extraction technology to extract parts information and assembly relation information, and realizes the transformation from assembly model to virtual assembly planning model. According to the modularization method, the methods and algorithms of key data planning, automatic assembly, assembly sequence planning and dynamic simulation of assembly process are studied, respectively.

Chapter 6 proposes a virtual assembly human–computer interaction method, to solve the position tracking Glove Ultra Polhemus and force feedback device, data gloves phantom a desktop, and other interactive devices on Win32 platform interface and drives, and on the application of the series of human–computer interaction hardware-related technology are studied, force sensing, touch successfully applied into the system, users can more deeply perceive the internal structure of machine equipment and the actual running state.

Chapter 7 proposes a virtual assembly network method is proposed. According to the characteristics of two different user groups, enterprises and the public, two systems for internal enterprises and public services are set up, respectively, which solve the urgent problem of remote access to virtual reality resources.

Chapter 8 puts forward the virtual single machine simulation operation method and technology, aiming at the shearer, scraper conveyor and hydraulic support single machine, using the established physical information sensing system, find the single machine working condition monitoring method. Then, under Unity3D environment, the single machine virtual simulation method of Three machines in fully mechanized coal-mining face which is identical with the reality is studied, including the seamless linkage method of hydraulic support components, virtual human-computer interaction mode of virtual hands, virtual bending technology of scraper conveyor and virtual memory cutting method of shearer.

Chapter 9 proposes a virtual collaborative simulation running method and technology, using the physical simulation of virtual reality software Unity3D engine, between the virtual equipment and equipment and the interaction relationship between virtual coal seam to compile, for the operation of the key information, characteristic and status, and the purpose is to realize the "+" equipment of coal seam joint virtual simulation run. A virtual coal seam simulation system composed of virtual inherent coal seam and virtual real-time update coal seam was established based on reverse reconstruction technology and Mesh collider. The equipment was co-simulated, the virtual interaction model between equipment and coal seam was established, and the simulation method of equipment behavior directly acting on coal seam floor was studied.

Chapter 10 introduces the virtual planning operation technology. From the perspective of artificial intelligence and VR simulation, based on the three machines in fully mechanized mining face collaborative mathematical model of fully mechanized mining and MAS theory, a VR collaborative planning method based on MAS is proposed under the Unity3D simulation engine, and a prototype system (FMUNITYSIM) is established. The key parameters of three machines in fully mechanized mining face can be planned and adjusted online, which provides a theoretical basis for fast planning and safe production of fully mechanized mining face.

Chapter 11 introduces sensor information construction technology and virtually real interaction technology. Sensor is arranged at the key position reflecting the pose information of equipment. Through a series of interfaces and channels, information interaction between Unity3D and configuration software, database and computing software is constructed. The distributed collaborative driving mode is adopted to optimize the data transmission and processing. Finally, the real-time operation data of fully mechanized mining equipment is imported into the virtual platform to drive the operation of virtual equipment, thus achieving stable and reliable sensing information collaboration and scheduling. Through the arrangement of wired/wireless full coverage network communication platform and the installation of digital sensing components on the equipment and intelligent transformation, the interconnection and interconnection of fully mechanized mining equipment are realized. Through the deep integration of VR monitoring system with data and video monitoring system, the real-time transparent presentation and remote intelligent coordinated control of the operating conditions of fully mechanized mining equipment can be realized.

Chapter 12 studies the condition monitoring and virtual simulation method, about "compressors" in actual working condition of the fully mechanized coal face, the constraints of the connection relationship between, in turn, the attitude behavior of the composite study, including the ideal floor flat cases "compressors" virtual collaborative technology, coal winning machine and the coupling between the scraper conveyer under different environment—coal winning machine and scraper conveyor of the posture under bending section of the feed coupling, coal winning machine and scraper conveyor under complicated working condition of joint positioning of informant method and the mutual influence between group of hydraulic support posture—group of hydraulic support memory monitoring method.

Chapter 13 introduces virtual monitoring system. Firstly, the integration of the theory of "Digital Twin" and the design of equipment in fully-mechanized mining face is analyzed. Then the system is designed as a whole. And then, six key technologies such as variable reservation and virtual interface in VR monitoring method based on Unity3D, as well as VR monitoring and real-time synchronization method based on LAN, are studied to provide software technical support for VR monitoring. Finally, the prototype system is designed and introduced.

Chapter 14 summarizes the conclusions of the book.

Chapters 1, 2, 4, 8, 9, 11 and 14 were written by Xuewen Wang, Chaps. 3, 5, 6, 7 and 12 by Jiacheng Xie, Chaps. 10 and 13 by Suhua Li. The research results and publication of this book are supported by projects such as the National Natural Science Foundation of China: "Precise virtual reconstruction of real-time position and attitude of a scraper conveyor and hydraulic support groups in complex floor environments" (52004174), Key Research and Development Plan of Shanxi Province: "Virtual reality based "Man-Machine-Loop" hardware in the loop simulation system for fully mechanized coal mining equipment" (201903D121141) and the Applied Basic Research Plan of Shanxi Province: "Research on virtual cooperative operation method of fully mechanized coal mining units in complex coal seam environment" (201901D211022).

Due to the author's limited knowledge level, there are some defects in the book, so we sincerely ask the readers to criticize and correct them.

Taiyuan, China
May 2021

Xuewen Wang
Jiacheng Xie
Suhua Li

Contents

Chapter 1
Application Overview

1.1 Introduction

Under the background of "Industry 4.0" and "Internet+ ", new generation of information technology, including cloud computing, mixed reality, big data, Internet of Things and Cyber-Physical Systems (CPS), has gradually begun to merge deeply with modern industry. All these urgently require unmanned intelligent mining technology to be strengthened and accelerated.

The digital twin relies on industrial simulation and visualization software technology, which can transform the various modules in the physical heaving workface into data integrated into a virtual system, realizing the simulation and analysis of each work process, work unit, and elements of fully mechanized mining operations in real time, accurately, and closer and more detailed. The simulation and analysis are based on historical data and combined with the actual operation of fully mechanized mining operations. In this system, each work and process in the fully mechanized mining operation is simulated and implemented, and various interactions are realized with it.

Virtual reality simulation technology is an important software and platform support for these efforts. The 'Made in China 2025' technology roadmap for key areas lists VR as one of the key technologies for core information equipment for intelligent manufacturing, and VR + industry helps accelerate the implementation of digital twin technology. Previous research work has been done in the area of fully mechanized mining, including virtual reality display, scenario simulation, training and education, and virtual monitoring and diagnosis, providing a good basis for the development of fully mechanized mining digitization. Overall, the digital design capabilities of fully mechanized mining are more uneven and still require integration, reform and continued innovation to clarify the next steps in development.

X. Wang et al., *Virtual Reality Technology in Mining Machinery*,
https://doi.org/10.1007/978-981-16-4408-5_1

1.2 Research Developments in Virtual Reality Assembly Technology and Systems

1.2.1 Virtual Assembly

Virtual assembly technology is an important application of virtual reality technology in engineering and is one of the key pieces of knowledge to be acquired in this book.

At present, manual assembly planning of large equipment assembly models is a rather complex and tedious task. In a particular series of equipment, the structure of most of the assembled parts and the structure of the selected components have similar characteristics. Using the secondary development tools provided by CAD software and the collaborative virtual assembly technology to develop the virtual assembly system, the application program extracts the characteristics of similar series of assembly parts, and then carries out the automatic assembly planning, which will greatly save the workload of repeated operation of designers, and improve the efficiency of R&D and design.

A virtual assembly system called Archimedes has been developed by those involved at Sandia National Laboratories in the USA [1]. The system has been used in actual production at NASA, Lockheed Martin, Rockwell and Hughes Aircraft, among others, with a well-known example being the Hughes Aircraft product, the 471-part fourth-generation air-to-air missile "Rattlesnake" AIM-9X.

1.2.2 Coal Mine Scenario Simulation

Internationally, developed mining countries such as Canada, the United States, Australia, and the United Kingdom have a relatively high degree of mine informatization, and the application of virtual reality in mines started early. The main research situation is as follows: The Australian Institute of Science and Industry (CSIRO) has established a fully mechanized mining working face training system [2]; Orr in the United States designed a mine escape system that can train four miners at the same time [3]; Dr. Jennifer Tichon and others established evaluation criteria for judging the current mainstream coal mine safety training system, and Experimental certification has been carried out to prove that the coal mine safety training system is indeed of great significance for improving the safety awareness of miners [4]. Canada's MARARCO Company has established a virtual reality laboratory VRL for mining engineering, which can realize all aspects of mining engineering [5].

In China, virtual reality technology has also started to be widely used in mining engineering, with some Chinese universities and others using virtual miners to actively explored typical accidents in underground coal mines, such as water penetration, fire, gas explosion and other accidents, or the scene simulation of fully mechanized mining face or fully mechanized excavation face.

1.2.3 Virtual Reality Interactive Tools

In the virtual reality system, the operator obtains various sensory stimuli and then obtains key information of the virtual object mainly through visual, auditory, force and touch sensing devices. Among them, virtual hand and haptic interaction technology play a pivotal role in the interaction between the operator and the virtual environment.

Abroad, Ritchie et al. realized a virtual assembly system based on force haptics [6]; researchers from the Georgia Institute of Technology in the United States developed a force feedback assembly-based human–computer interaction system, HIDRA, using two phantom devices, and after testing, obtained a method to realize collision interference inspection and force haptic feedback functions using phantom [7]; in the research of human–computer interaction in virtual assembly, the University of North Carolina used phantom device to combine CAD oriented modeling technology with actual manufacturing to summarize the feasibility of product design [8]; researchers in the Netherlands and Spain have also used phantom devices to do related research on data synchronization, information processing and assembly simulation [9, 10].

The design and development of domestic force feedback equipment mainly began in the early twenty-first century. Among them, the research trends of phantom series force feedback devices are as follows:

In terms of industry, some universities in China have also conducted some research using virtual reality interactive tools, including virtual scenario of a ship, robot remote control operating system and virtual assembly technology based on force/tactile feedback.

1.2.4 3D Modelling of Coal Mining Equipment

Modelling is the creation of the mathematical structure of an object based on the intrinsic laws of a specific object in the objective world, after a reasonable simplification of the hypothetical processing, with the help of the right mathematical tools to express the interrelationships of real objects. Similarly, the ability to correctly and rationally model the behavior of solid objects and to present a realistic virtual world is crucial in today's era of rapid development of networked virtualization.

Some universities in China make use of the advantages of virtual system, such as immersion, interactivity, authenticity and easy realization, build the model with the help of relevant 3D rendering software and modeling software, develop the virtual simulation system through visual programming technology, and establish a comprehensive prediction, early warning and visual mine mining virtual system.

With the continuous development of Web3D technology, the methods and means of modelling technology are also being enhanced. For example, WebGL technology can create 3D models and interactive 3D animations directly through the HTML script itself or import external models through the loader, and then achieve graphic rendering through the OpenGL interface, truly realizing the display and interaction of mechanical 3D models on the browser side. Moreover, it has become the general trend of 3D display technology today because it does not need to install any rendering plug-ins on the browser side.

1.2.5 Web-Based Information Service Platform and Virtual Assembly Technology

In recent years, the establishment of a network-based information service platform for a certain engineering field has become a hot spot for research in the field of computer-aided engineering. Many scholars have made active exploration in this field [11], established a framework of common key technology system for cloud manufacturing service platform for small and medium enterprises, which laid the foundation for the network-based service platform research foundation [12]. In the field of coal, a series of explorations using advanced technologies such as the Internet of Things and cloud computing [13–16] have also provided a new means and technical guarantee for coal production, management and safety.

In the field of coal mining equipment, a cloud manufacturing resource service platform for coal mining equipment has been established [17], which can provide resource services, data services, computing services, software applications and technical support for coal mining equipment design and application units; The CAE service system for mining machinery and equipment [18] can provide manufacturing enterprises with remote CAD/CAE rapid parametric design and finite element analysis services.

As network technology continues to mature, network virtual assembly is gradually being developed. If it is applied to remote presentation and training, and at the same time with the help of advanced 3D virtual simulation technology, inevitably endow distance education with more profound interest, flexibility and interactivity, and it has great development potential.

1.2.6 Web-Based Digital Modelling and 3D Dynamic Design

Today, the network has been fully integrated into every field of human daily life, at the same time, customers are also improving their requirements, especially the interactive performance, transmission performance, rendering quality and so on. The emergence of Web3D technology can fully demonstrate its powerful ability in generating interactive 3D graphic objects, and has been widely used in the field of 3D product display of Internet e-commerce.

At present, some scientific research institutions use the relevant 3D modeling technology and display technology to study the realization of real-time and interaction in the virtual environment, realize the rapid prototyping of the model in the virtual environment and the cross platform operation of virtual development, enhance the immersion of the operator, and provide ideas for industrial production.

1.2.7 Design of Public Service Platform for Mechanical Equipment

With the rapid development of network and digital design technology, the network has evolved from a media of information dissemination to a platform for information computing and sharing. According to the service objects of the platform, the platform is generally divided into public service platforms of science and technology, innovation, education and intellectual property information. How to establish a public service platform which includes all kinds of important resources and can realize the sharing of various infrastructure equipment has gradually become an indispensable technical support for enterprises.

The field of coal mining mechanical equipment has also gradually begun to use the Internet of Things and cloud computing to present some infrastructure, equipment, data resources, software resources, IT resources, information resources and so on to users in the form of services through the Internet, providing powerful technical support for safe coal mining and management [19]. For example, the successful application of cloud manufacturing resource service platform of coal mining mechanical equipment and the collaborative development of service-based manufacturing in cloud computing environment can provide knowledge resources, calculation, analysis and software sharing and technical support for the manufacturing enterprises and users related to coal mining mechanical equipment, which can significantly improve the utilization rate of resources and can improve the quality and efficiency of coal mining mechanical equipment R&D and design.

Building a digital model integration platform for coal mining mechanical equipment with commonweal, fundamentality and independent intellectual property rights and using an adaptive operation mechanism, can provide remote structural design, assembly process and operation scenarios for coal mining mechanical equipment, and can provide technical and resource support for manufacturing enterprises related to coal mining mechanical equipment and users of coal mining mechanical equipment, which has important value. At the same time, it can provide a unified auxiliary solution for such user groups, and ultimately achieve the purpose of improving resource utilization and strengthening public information sharing.

1.3 Research and Development of Monitoring and Dynamic Planning in Fully Mechanized Coal-Mining Face Under Virtual Reality Environment

1.3.1 The Monitoring Method in Fully Mechanized Coal Mining Face

1.3.1.1 Positioning and Attitude Determination Method of Shearer

(1) State information collection and monitoring method of shearer
In terms of the research on state information collection and monitoring method of Shearer, JNA crossheading system (JOS) of Joy Company realizes the automatic control of various equipment in fully mechanized coal-mining and the collection and uploading of real-time data. The IPC control system of Eickhoff company in Germany realizes the collection, processing and storage of the status information of the shearer, displays the operation state of the shearer in various ways such as text, data, curve and graph, and transmits it to the crossheading and ground by monitoring network.

In China, the PLC display carried by the electric traction shearer is commonly used in the underground mechanized coal-mining face to monitor the current, voltage, speed, temperature and fault display of the shearer when it is running. Operators operate on site, and the monitoring screen of upper machine is not intuitive, which brings inconvenience to the fault analysis and overall management of shearer [20].

(2) Shearer positioning method
For the research of shearer positioning at home and abroad, the traditional methods mainly include gear counting method, infrared radiation method and ultrasonic reflection method, which have been widely used in the actual production. But all these methods have disadvantages such as accumulated errors and inability to continuously monitor [21].

Inertial Navigation System (INS) is an autonomous navigation system with the advantages of high data update rate, comprehensive data and high accuracy of short time positioning [22]. It is of great significance to apply inertial navigation system to the shearer equipment and monitor the whole mining process.

The Strapdown Inertial Navigation System (SINS) is more optimized and less costly than INS, but has the same drawbacks as INS. The biggest shortcoming in the application of shearer positioning is that the distance of the underground roadway is too short so that some errors will appear. It is necessary to gradually eliminate the errors through calculation.

The main problem with the shearer positioning is that if a single positioning method is used for the shearer positioning, a certain error will always be generated.

(3) Monitoring method of shearer attitude

Nowadays, some fully mechanized coal-mining faces use the inclination angle of the shearer to detect the change of terrain and the range of undulations. In the memory cutting, the shearer lateral and longitudinal inclination angle data obtained in real time are used to compensate the memory cutting height data [23]. However, the inclination angle of the shearer body is the fluctuation between the front and rear support slippers (Guide slippers). The distance between these two slippers often has the length of four to six middle troughs, and the coal seam changes between the slippers may be ignored, showing an 'insensitivity' to terrain changes, making this approach unreliable.

In general, the height of the roof and floor along the working face changes slowly, and abrupt changes occur only when faults appear. In either case, at the same sampling point on the working face, the change of the floor is generally not obvious between adjacent cutting cycles. However, the accumulation of gradual changes in the roof and floor, errors in the positioning of the shearer between adjacent cutting cycles, and the accumulation of the gradual change in the attitude of the shearer all have an impact on the attitude monitoring and memory cutting of the shearer.

1.3.1.2 Hydraulic Support Monitoring

Hydraulic support is the key support equipment for fully mechanized coal-mining face in underground coal mines. Its operating status, including support pressure and attitude, directly affects whether the entire working face can be mined and worked in a safe and efficient way, so it is vital to keep track of the attitude of the hydraulic supports [24]. In the actual working face, the number of hydraulic supports is more than one hundred. Although each hydraulic support operates independently, all the seemingly separated individuals can complete the roof support task together under the control of a certain operating law [25]. Therefore, in the monitoring process, it is necessary to monitor the hydraulic supports from an overall point of view.

The monitoring of hydraulic support mainly focuses on pressure monitoring and attitude monitoring. At present, the method of attitude monitoring for hydraulic support is mainly to monitor the key parameters of the single hydraulic support. The general process of attitude monitoring of hydraulic support is to install sensors in the reasonable position of hydraulic support, transmit the collected data to the upper computer through wireless or wired network, and judge whether the data is within the reasonable threshold range in real time to determine whether the work is normal, and create appropriate curves and alarm reports in the screen of monitoring software.

1.3.1.3 Scraper Conveyor Monitoring

(1) Scraper conveyor attitude monitoring
In the aspect of the research on the attitude monitoring of scraper conveyor, Jiang et al. [26] designed a scraper conveyor attitude control system and control method based on wireless 3D gyroscope technology to achieve attitude monitoring of scraper conveyors. Wu [27] used the inclination sensors and potentiometers in the side plates of each middle trough of the scraper conveyor to obtain an accurate trend of the state change of the bottom plate of the working face.

(2) Current research status of scraper conveyor bending section monitoring
The length and angle of the scraper conveyor's bending degree are determined by the telescopic length of the pushing oil cylinder of the hydraulic support corresponding to each chute, which directly affects the movement track of the shearer's feed, and further directly determines the stability and reliability of the shearer's operation. Usually, the research on scraper conveyor is about the meshing between sprocket and chain, dynamics and friction and wear under the condition of non-bending [28]. The research on bending section focuses on the analysis under the premise of assuming the bending section shape [29]. However, there are few studies on the accurate solution of the actual posture of the curved section of the scraper conveyor [30].

1.3.2 VR Scene Simulation of "Three Machines" in Fully Mechanized Coal-Mining Face

VR technology has been widely used in the field of coal mine training and teaching because of its intuitive, immersive and interactive characteristics, and has achieved great application value [31].

Tichon et al. [32] applied VR to miners' safety training exercises, effectively improving miners' safety awareness. Kerridge et al. [33] established a framework for evaluation, and used VR technology to analyze and evaluate underground risks. Foster et al. [34] studied the interactive immersion simulation operation training of underground miners. With the help of head-mounted display (HMD) and other interactive means, the remote operation training of underground continuous shearer and drilling machine was carried out respectively.

Based on complete simulation data, Akkoyun [35] established an interactive visualization environment for teaching and learning related to mining engineering to show the production process and conditions of open-pit magnesite.

At present, the methods to realize VR simulation of fully mechanized coal-mining face are mostly based on VR simulation development engine represented by Unity3D, Quest3D, and OSG. Combined with 3D modeling software such as UG, Pro/E and 3D MAX, the motion programming of virtual equipment and the design of GUI interface are carried out, and then the virtual model is interactively controlled to realize the virtual simulation of fully mechanized coal-mining face, and then the application of training and teaching is carried out.

1.3.3 VR Monitoring of "Three Machines" in Fully Mechanized Coal-Mining Face

VR technology has made many achievements in training and education in the field of coal mines. In recent years, some scholars [36, 37] have gradually applied VR technology to the field of fully mechanized coal-mining monitoring, and carried out real-time, reliable, visual, and pre-operated remote monitoring and control [38]. It can not only control the accurate operation state of fully mechanized coal-mining equipment in time, but also predict its posture and state, so as to carry out fault prediction, adjust operating parameters in a timely manner, and carry out necessary artificial remote intervention [39, 40], so as to realize on-site remote control and scheduling. At present, the technical means for VR monitoring of fully mechanized coal-mining equipment are mainly based on VR platforms such as Virtools, EON and Unity3D for design, integration and innovation.

(1) Based on Virtools platform and method
Virtools software allows users to quickly and simply implement 3D interactive applications through editing the behavior module. In China, the earliest research on VR monitoring of fully mechanized mining face mainly uses this software. Since the software interface is friendly and easy to start, it has been favored by many scholars. They basically connect Virtools with ready-made interfaces of SQL SERVER. In addition, the number of sensor interfaces used for monitoring is small. Most users are engaged in theoretical exploration of VR monitoring technology and have not entered the industrial test stage. However, due to the limited network function of Virtool software and the stop of updating and upgrading many years ago, the technology is gradually lagging behind. It is only suitable for win7 and below platforms and cannot keep up with the progress and development of VR technology, so it is gradually eliminated.

(2) Based on EON platform and method
EON software provides a new set of modules with excellent performance, realizing industrial-level VR design. Since the research of EON software for fully mechanized coal-mining VR monitoring mainly arose after Virtools software was gradually phased out, the operation methods were similar, and it was easy to get started. For a while, it became a popular software used by many scholars. However, its limitations are also obvious. EON software is inconvenient to connect to databases such as SQL SERVER and cannot be used for data storage. In particular, there are serious defects in organizing large-scale scene production, so it cannot be applied to the VR monitoring research of large and complex scenes of fully mechanized coal-mining face.

(3) Based on Unity3D platform and method
Unity3D is a professional VR simulation engine, and its introduction into the coal mine field is of great significance for improving coal mine automation and visualization. Based on the superiority of Unity3D software in VR monitoring, virtual technology has gradually transited from single machine research to collaborative monitoring and research stage of Three machines in fully mechanized mining face, which realizes the synchronization of data and virtual state, the smooth connection and reading of database. However, the research is only in the initial stage, and does not consider the particularity and complexity of Three machines in fully mechanized mining face working condition monitoring under actual working conditions, so its ideal state and theoretical color are relatively strong.

(4) Other methods
In addition to Virtools, EON and Unity3D software, some scholars tried to use other software to explore the VR monitoring of fully mechanized coal-mining, such as Quest3D and WPF.

1.3.4 VR Planning of "Three Machines" in Fully Mechanized Coal-Mining Face

Since the Three machines in fully mechanized mining face coordination control program has the static and single parameter setting, and the program, process and parameters are also static and single, the program cannot adapt to the complex and changeable working face environment and is prone to unknown errors and problems, so the complete set of intelligent collaborative control of fully mechanized coal-mining cannot become the mainstream production mode of mining operation in China. In order for the Three machines in fully mechanized mining face to run adaptively and dynamically in a dynamically changing underground environment, it is necessary to have the ability of "living body", self-awareness, self-decision-making, and self-adaptation.

In Industry 4.0, Digital Twin refers to copying a physical object in a digital way, simulating the behavior of the object in a real environment, realizing the virtualization and digitization of the whole process, so as to solve past problems or accurately predict the future. Siemens carries out the design and application of unmanned chemical plants, which can digitize all elements of the actual processing and production process. Afterwards, at the digital level, all the links in the production process are analyzed through simulation, and it can also predict what will happen after the actual production, so as to prevent the investment loss caused by many problems in the actual production or production process. At present, the digital design ability of VR research in the field of coal mining mechanical equipment is still relatively low, and VR technology cannot carry out comprehensive virtual planning of fully mechanized coal-mining face.

1.4 The Structure and Content Arrangement of the Book

This book is based on the research background of coal mining mechanical equipment in fully mechanized coal mining face. It conducts research from three aspects of virtual assembly, virtual planning and virtual monitoring of coal mining mechanical equipment in a virtual environment. The book is mainly divided into the following parts.

First of all, this book focuses on the overall design of the system, introduces the composition and operation process of the production system of the fully mechanized working face based on digital twinning, and introduces in detail the key technologies of each stage of the production system and the related cases completed under laboratory conditions.

Secondly, this paper summarizes the transformation technology of virtual simulation element model and the establishment process of model library. According to the different characteristics of the structure and actual work of shearer, scraper conveyor, road header and hoist, the structure is divided in detail and the virtual reality model resource library is established.

Two virtual assembly methods are introduced, which are OSG-based virtual assembly method and UG-based virtual assembly method. The conversion from assembly model to virtual assembly planning model is realized. The functions of key data planning, automatic assembly, and assembly sequence planning and assembly process dynamic simulation can meet the requirements of users for virtual assembly.

In this book, we propose a solution to the application of a variety of virtual reality human–computer interaction hardware devices in this system, and study the interface and driving problems of a variety of human–computer interaction devices, such as position tracker GloveUltra, data glove Polhemus and force feedback phantom desktop. At the same time, we study a series of related technologies of human–computer interaction hardware applications. Based on the operation mode of web-based virtual assembly and simulation system, according to the characteristics of two different user groups of enterprises and the public, two kinds of systems for internal enterprises and public services are built respectively.

Based on the realization of virtual assembly, in the VR environment, taking the Three machines in fully mechanized mining face equipment of fully mechanized working face shearer, scraper conveyor and hydraulic support as the research object, in order to establish the virtual image of the real Three machines in fully mechanized mining face real-time working operation state, the VR monitoring and planning method is studied, trying to establish a VR monitoring and planning system with high reliability, strong real-time performance, and panoramic 3D display of the operating conditions of the fully mechanized coal-mining face. This book mainly studies the virtual planning and virtual monitoring from the aspects of virtual single machine simulation operation method and technology, virtual collaborative simulation operation method and technology, virtual planning operation technology, sensor information construction technology and virtual real interaction technology, condition monitoring and virtual simulation method, and the establishment of virtual monitoring system.

References

1. Ralston, J., Reid, D., Hargrave, C., et al.: Sensing for advancing mining automation capability: a review of underground automation technology development. Int. J. Min. Sci. Techno. **3**(24), 19–24 (2014)
2. Mallett, L., Unger, R.: Virtual reality in mine training. Soc. Min. Metal. Explor. **59**(02), 1–4 (2007)
3. Orr, T.J., Mallet, L.G., Klein, K.A.: Enhanced fire escape training for mine workers using virtual reality simulation. Min. Eng. **61**(11), 41 (2009)
4. Tichon, J., Burgess, L.R.A.: review of virtual reality as a medium for safety related training in mining. J. Health Safety Res. Pract. **3**(1), 33–40 (2011)
5. Foster, P.J., Burton, A.: Modeling potential sightline improvements to underground mining vehicles using virtual reality. Min. Technol. **115**(3), 85–90 (2006)
6. Ritchie, J.M., Lim, T., Medellin, H., et al.: A haptic based virtual assembly system for the generation of assembly process plans. Memorias Del XV Congreso Internacional Anual de La SOMIM, 585–593 (2009)
7. Luo, Q., Xiao, Q.: Physically accurate haptic rendering with dynamic effects. IEEE Comput. Graph. **24**(6), 60–69 (2004)
8. Strijp, C.J.V., Langen, H.H., Onosato, M.: The application of a haptic interface on microassembly. Symposium on Haptic Interfaces for Virtual Environment & Teleoperator Systems IEEE, **1**(1), 289–293 (2006)
9. Iglesias, R., Prada, R., Uribe, E., et al.: Assembly simulation on collaborative haptic virtual environments, pp. 241–247. Václav Skala Union Agency (2007)
10. Chen, H., Yuan, Y., Sun, W., et al.: The cloud manufacturing service platform based on the SLA of SMEs. Mod. Manuf. Eng. (07), 16–20 (2013)
11. Yin, C., Huang, B., Liu, F., et al.: Common key technology system of cloud manufacturing service platform for small and medium enterprises. Comput. Integr. Manuf. Syst. **17**(3), 495–503 (2011)
12. Qian, J.S., Shanshan, M.A., Sun, Y.J., et al.: Design on mine comprehensive automation system based on internet of things. Coal Sci Techno. **39**(02), 73–76 (2011)
13. Pei, Z., Li, B., Xu, S., et al.: Integrated platform system framework of mine internet of things based on cloud computation. Coal Sci. Technol. **40**(09), 90–94 (2012)
14. Sun, Y., Qian, J., Li, S., et al.: Theory and key technology for mine internet of things. Coal Sci Techno. **39**(02), 69–72+79 (2011)
15. Li, C., Mao, S., Jiang, Y., et al.: Design and implementation of mine virtual platform and electromechanical monitoring and measuring application. Coal Sci Techno. **39**(11), 94–99 (2011)
16. Wang, X., Yang, Z., Ding, H., et al.: Research and application on coal mine machinery equipment cloud manufacturing resource service. J. China Coal Soc. **38**(10), 1888–1893 (2013)
17. Yang, Z., Wang, X.: Study on cloud simulation CAE service system for mining machinery equipment. J. Mech. Eng. **49**(19), 111–118 (2013)
18. Hren, G.: Web-based environment for mechanism simulation integrated with CAD system. Eng Comput-Germany. **2**(26), 137–148 (2010)

19. Liu, X., Zhaolijuan, Gai, D., et al.: Research on intelligent hauling speed regulation system based on shearer's reliability. J. Syst. Simul. **28**(7), 1601–1608 (2016)
20. Ying, B., Li, W., Luo, C., et al.: Experimental study on combinative positioning system for shearer. Chin. J. Sens. Actuat. **28**(2), 260–264 (2015)
21. Hoflinger, F., Muller, J., Zhang, R., et al.: A Wireless micro inertial measurement unit (IMU). IEEE Trans. Instrum. Meas. **62**(9), 2583–2595 (2013)
22. Sammarco, J.J.: A guidance sensor for continuous mine haulage. Ias Meeting IEEE, **4**, 2465–2472 (1996)
23. Reid, D.C., Hainsworth, D.W., Ralston, J.C., et al.: Sheareer guidance:a major advance in longwall mining. Field Serv. Rob. **4**(24), 469–476 (2006)
24. Schnakenberg, G.H., Jr.: Progress toward a reduced exposure mining system. Min. Eng. **49**(2), 73–77 (1997)
25. Liu, C.: Mathematic principle for memory cutting on drum shearer. J. Heilongjiang Inst. Sci. Technol. **20**(2), 85–90 (2010)
26. Luo, S.H., Wu, Y.P., Xie, P.H., et al.: Mechanical analysis of support stability in longwall mining of steeply dipping seam. J. China. Coal. Soc. **44**(9), 2664–2672 (2019)
27. Verma, A.K., Deb, D.: Numerical analysis of an interaction between hydraulic-powered support and surrounding rock strata. Int. J. Geomech. **13**(2), 181–192 (2013)
28. Yang, Y., Gao, Z.: A New Method for Control Allocation of Aircraft Flight Control System. IEEE T Automat Contr. **65**(4), 1413–1428 (2020)
29. Wu, P., Niu, N.: Research and analysis on the relationship between scraper angle and working face state. Electron. World **16**, 98–99 (2012)
30. Mao, J., Shi, J., Zhang, D., et al.: Dynamic modeling and simulation of heavy scraper conveyor. J. China. Coal. Soc. **33**(01), 103–106 (2008)
31. Agnew, J.M., Landry, H., Piron, E.: Technical note: performances of conveying systems for manure spreaders and effects of hopper geometry on output flow. Appl. Eng. Agric. **21**(2), 159–166 (2005)
32. Liao, X., Zhang, J., Feng, T., et al.: Intensity nonlinear analysis of central trough of coal shearer under limiting conditions. J. Southeast. U: Nat. Sci. Ed. **44**(3), 531–537 (2014)
33. Mao, J., Shi, J.G. , Zhang, D.S., et al.: Dynamic modeling and simulation of heavy scraper conveyor. J. China. Coal. Soc. **33**(1), 103–106 (2008)
34. Mujber, T.S., Szecsi, T., Hashmi, M.S.J.: Virtual reality applications in manufacturing process simulation. Mater. Process. Tech. **155–156**(1–3), 1834–1838 (2004)
35. Tichon, J., Burgesslimerick, R.: A review of virtual reality as a medium for safety related training in mining. J. Health Safety Res. Pract. **3**(1), 33–40 (2011)
36. Naji, H.A.H., Xue, Q.J., Lyu, N.C., et al.: Evaluating the Driving Risk of Near-Crash Events Using a Mixed-Ordered Logit Model. Sustainability-Basel. **10**(8), 2868 (2018)
37. WoHrer, A., Brezany, P., Tjoa, A.M.: Novel mediator architectures for grid information systems. Future. Gener. Comp. Sy. **21**(1), 107–114 (2005)
38. Akkoyun, O., Careddu, N.: Mine simulation for educational purposes: a case study. Comput. Appl. Eng. Educ. **23**(2), 286–293 (2015)
39. Wang, J., Huang, L., Li, S., et al.: Development of intelligent technology and equipment in fully-mechanized coal mining face. J. China. Coal. Soc. **39**(8), 1418–1423 (2014)
40. Lee, W., Cho, S., Chu, P., et al.: Automatic agent generation for IoT-based smart house simulator. Neurocomputing **209**, 14–24 (2016)

Chapter 2
System Overall Design

2.1 Introduction

The concept of "smart mine" integrates the forward-looking and holistic thinking and research of many scholars in the digital mine and other related theories and application framework, and the state has also issued the "general technical specification for smart mine information system" and other related industry standards [1]. But the "smart mine" needs the support of various systems and fields in the mine, convergence and digital integration, as well as the realization of a large number of technological innovation and new ideas. The digitization of fully mechanized mining system is undoubtedly the top priority and primary goal of "smart mine". From the current aspects of digital design [2, 3], CAE analysis [4], virtual reality [5], motion monitoring [6] and underground Internet of things [7], although they have made rapid progress, they are still relatively independent and separated. How to effectively integrate them, and then accurately and carefully guide and lead the development of fully mechanized mining digitization, this is a difficult problem that must be solved first.

From the perspective of international and domestic frontier fields, digital twin technology [8], CPS [9] and other concepts have been integrated into the mechanical manufacturing production system, resulting in a series of research achievements such as digital twin Workshop [10], product digital twin [11]. The concept and technology of digital twin are more and more accepted by scholars. They are applied to digital twin of CNC machine tool [12], modular digital twin technology for plant planning [13], product management of whole life cycle [14], automatic production line of digital twin [15], after-sales service and support of products [16], nuclear energy equipment of digital twin [17], and building information model of digital twin (Building Information Modeling, BIM) [18], the distance between virtual space and real space is reduced, and good results are achieved. In the practice of unmanned factory, Siemens [19] analyzed all links in the production process through simulation, and realized the virtualization and digitization of the whole process from product design to manufacturing.

X. Wang et al., *Virtual Reality Technology in Mining Machinery*,
https://doi.org/10.1007/978-981-16-4408-5_2

Relying on the technology of industrial simulation and visualization software, digital twin can transform each module in the entity comprehensive working face into data and integrate it into a virtual system [20]. Through real-time interaction, human–computer interface, intelligent analysis and so on, it can realize simulation and analysis that is real-time, accurate, closer and more detailed integration with the actual operation of fully mechanized mining. These works are based on historical data. In this system, each work and process in the fully mechanized mining operation is simulated and interacted with.

The underground environment and conditions are bad, the space is closed and narrow, and the geographical environment is unpredictable. Therefore, there may be danger all the time, and the equipment operation may fail at any time, which determines that the production system of fully mechanized mining face based on digital twin is much more challenging than the ground production system such as digital twin workshop. In this paper, the overall model, operation process and key technology of production system design and operation mode of comprehensive working face based on digital twin are studied in detail, and the development trend of various technical fields is studied, trying to find out which technical achievements are in place and which technologies are still in great shortage, so as to analyze the reasons, find the gap and speed up the catch-up.

2.2 Overall Model and Operation Process of Design and Operation Mode of Comprehensive Working Face Production System Based on Digital Twin

The production system of comprehensive working face is reproduced in the virtual environment, so that the virtual space model and physical space model of the production system of comprehensive working face are in real-time interaction, and can respond to each other's dynamic changes in time [8]. The digital twin technology is used to design fully mechanized mining equipment products, coal seam geological information conditions, equipment selection and process arrangement, equipment layout and virtual operation realize seamless integration, improvement and optimization of real-time monitoring and other processes and production system.

2.2.1 Overall Model

As shown in Fig. 2.1, the Digital Twin Fully Mechanized Coal Mining Face (DTFM) production system is mainly composed of Physical Fully Mechanized Coal Mining Face (PFM), Virtual Fully Mechanized Coal Mining Face (VFM) and Fully Mechanized Coal Mining Face Design and Monitoring System (FMDMS) and Fully Mechanized Coal Mining Face Digital Twin Data (FMDTD).The specific relationship

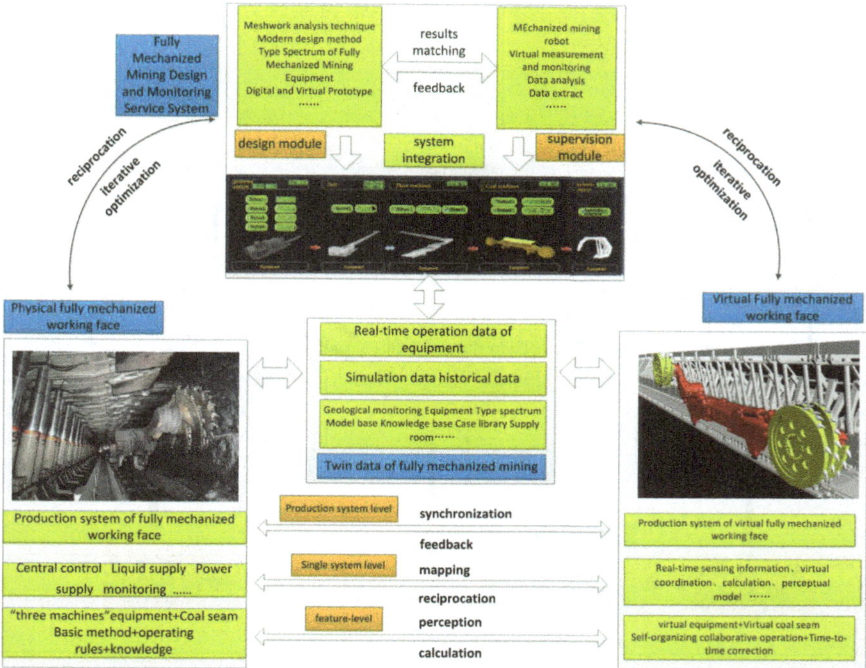

Fig. 2.1 General model

is: Based on the design theory method and monitoring bottom operation model provided by the design module and monitoring module of FMDMS, through the two-way real mapping and real-time interaction of PFM and VFM at the element level, single system level and production system level, as well as the real-time update of knowledge base, historical data and equipment real-time operation data provided by FMDTD, the complete integration of PFM, VFM and FMDMS is realized integration and fusion of elements, whole process and whole data. Driven by FMDTD, the iterative operation of fully mechanized coal mining face's production factor management, pre-simulation and real-time control of production process in PFM, VFM and FMDMS is realized, so as to achieve the goal of safe and efficient mining of production system configuration and equipment cooperation in comprehensive working face under the constraint of specific geological conditions.

2.2.2 Operation Process

The operation process is shown in Fig. 2.2. Stage 1 is the iterative optimization process of face element management, which reflects the interactive process of PFM and FMDMS in DTFM, in which FMDMS plays a leading role. When DTFM receives

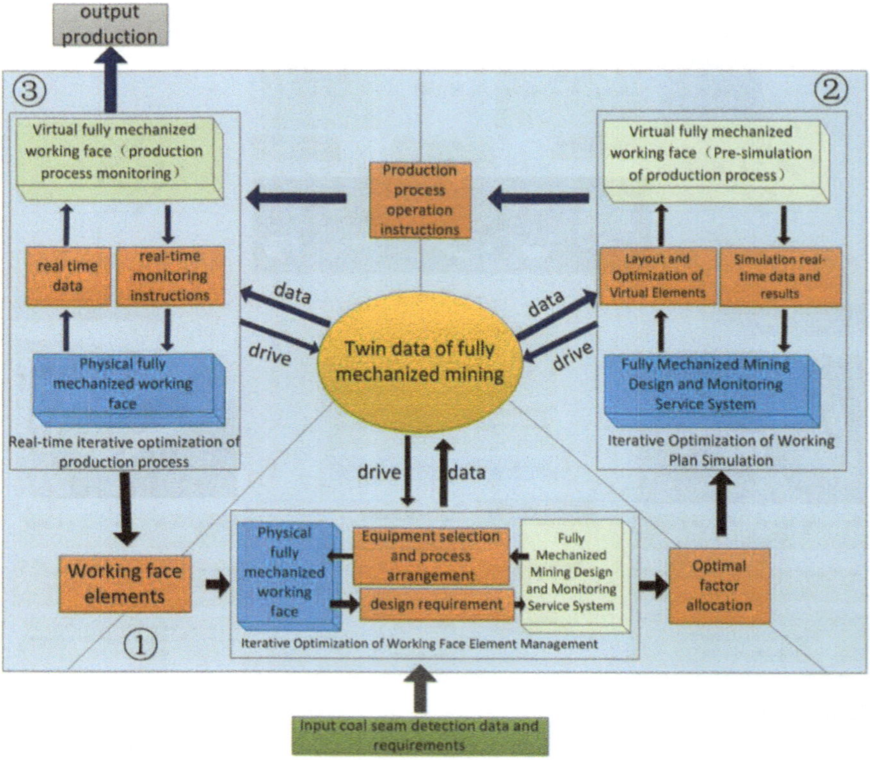

Fig. 2.2 Operation flow of digital twin fully mechanized mining face

a production task (mainly coal seam detection data and related conditions and parameters, etc.), the design module of FMDMS, driven by the historical data and other related data in the corresponding fully mechanized coal seam and equipment selection database, analyzes, evaluates and forecasts the production factors, and obtains the initial configuration project of equipment selection and layout, coal mining technology and process to guide the design of PFM.PFM sends the real-time data of the actual layout process of the working face elements to FMDMS for model reasoning, case-based reasoning and optimization, and guides the corresponding correction of the working face elements. After repeated iterations, the optimal allocation of working face elements is obtained.

Phase 2 is the iterative optimization process of production planning, which reflects the interaction process between FMDMS and VFM in DTFM, in which VFM plays a leading role. VFM receives the configuration of the initial optimal working face elements generated in the first stage, generates the virtual working face which is completely consistent with PFM, establishes the virtual models of equipment and coal seam, and arranges the equipment in the virtual coal seam, giving it the corresponding operation behavior. Driven by the historical simulation data, simulation real-time data and other related data in FMDTD, based on various elements, behavior and rule models, and simulating various disturbances and problems encountered in the actual operation, the fully mechanized production plan is simulated, analyzed and optimized. The simulation results are sent to the FMDMS to revise and optimize the production plan and return to the VFM. After repeated iterations, the optimal production plan is obtained and the production process operation instructions are generated.

Stage 3 is a real-time iterative optimization process of production process, which reflects the interaction process between PFM and VFM in DTFM, in which PFM plays a leading role. PFM carries out the actual production according to the production process operation instructions generated in stage 2, and transmits the real-time data to VFM. VFM compares the actual operation data with the pre-defined production plan data, and inputs the actual disturbance factors to continue the virtual simulation. It evaluates, optimizes and forecasts the production process, finds out the possible problems in the operation, and input the production operation optimization instructions into the centralized control system in the form of program and acted on PFM in the form of real-time control instructions to optimize the production process. Such iteration is repeated until the optimal production process is achieved.

Under the condition of our laboratory, a design prototype system of digital twin fully mechanized coal mining is preliminarily realized. The three stages can respectively complete the functions of digital design of fully mechanized coal mining equipment, virtual simulation and optimization of fully mechanized coal mining production system, and real-time virtual monitoring and control of fully mechanized coal mining production system.

2.3 The First Stage: The Theory and Method of Production System Design of Fully Mechanized Working Face

2.3.1 Carry on the Design of Full Life Cycle of Fully Mechanized Mining Equipment Products

The design of fully mechanized mining equipment should refer to the design process of general mechanical products, greatly improve the digital design level of products, establish the digital prototype and virtual prototype throughout the whole life cycle, which are suitable for each production stage, and carry out collaborative design. In the design process, it is also necessary to combine with the actual operating conditions of the products in the underground, and carry out specific design for some special needs of fully mechanized mining equipment.

The specific process is to establish a series of fully mechanized mining equipment products, and combine the Internet, database and modern design theory methods to realize the digitization, networking, intelligence and integration collaboration of the design process of fully mechanized mining equipment products, such as conceptual design, parametric modeling, virtual assembly, CAE analysis, optimization design, reliability design and knowledge management.

The knowledge base of fully mechanized mining equipment products is established, including case base, parts base, material base, CAE analysis knowledge base, etc. the mixed knowledge representation model is used to realize the representation and encapsulation of design cases, rules and design process, and effectively manage the design knowledge of fully mechanized mining equipment products.

Break through the common key technologies of the design and manufacturing of fully mechanized mining equipment, including the collaborative call technology of various commercial design software (Matlab, ANSYS, UG, Adams) under the network environment, the virtual prototype modeling technology of the whole machine and key parts, the dynamic, static, rigid flexible hybrid analysis technology, the structural strength, optimization design, structural modal and fatigue life analysis technology.

Finally, the design digitization, resource integration, operation networking, management informatization, service intellectualization and process automation of fully mechanized mining equipment can be realized, which can shorten the product development cycle, reduce the product development cost, improve the product quality and enhance the product market competitiveness.

2.3.2 *Coal Mine Enterprise Selection and Configuration Design*

Based on the design of fully mechanized mining equipment products, the selection and configuration design of fully mechanized mining production system is carried out. It mainly includes:

- The common type spectrum of fully mechanized mining equipment is established, and a series of virtual models and behavior database of production process are established. It includes equipment characteristic size database, equipment performance parameter database, equipment characteristic simulation script database, equipment characteristic parts parametric model database, equipment whole machine assembly model database, and constructs digital virtual equipment which can simulate the real behavior of equipment.
- In order to better serve the design process of fully mechanized coal mining, the multi-level knowledge model of fully mechanized coal mining production system is constructed, and the knowledge base of key elements of working face is established. The object of establishment mainly includes "human–machine–environment", among which "human" refers to the operating object that can control the equipment, which is a narrow concept, and needs to consider the automation level of the control system and the subjective initiative of the operator; "machine" mainly refers to the fully mechanized mining equipment, including shearer, scraper conveyor, hydraulic support, hydraulic system and other fully

mechanized mining equipment and its behavior; "environment" includes coal seam, surrounding rock, mine pressure, roof breakage and various disasters (water, fire and gas).

- According to the data, the virtual coal seam is constructed, which is driven by real-time geological data. The feature data of coal seam surface can be extracted and transformed into the cutting path that the equipment can run.
- Based on the above three points, an integrated software tool of intelligent comprehensive mining cross domain design and simulation is constructed to realize the optimal allocation of working face elements efficiently and quickly.

2.3.3 *Case*

In the first stage, the design schematic diagram is shown in Fig. 2.3, the digital design system of fully mechanized mining equipment is established, which is composed of eight subsystems: conceptual design subsystem, CAD parametric modeling subsystem, virtual assembly subsystem, CAE analysis subsystem, design service subsystem, knowledge management subsystem, literature and training subsystem, and other auxiliary function subsystem. What's more, case management system, rule base, material base, model base and other knowledge bases are established, and the data of each link can be transferred to each other, so as to complete the whole

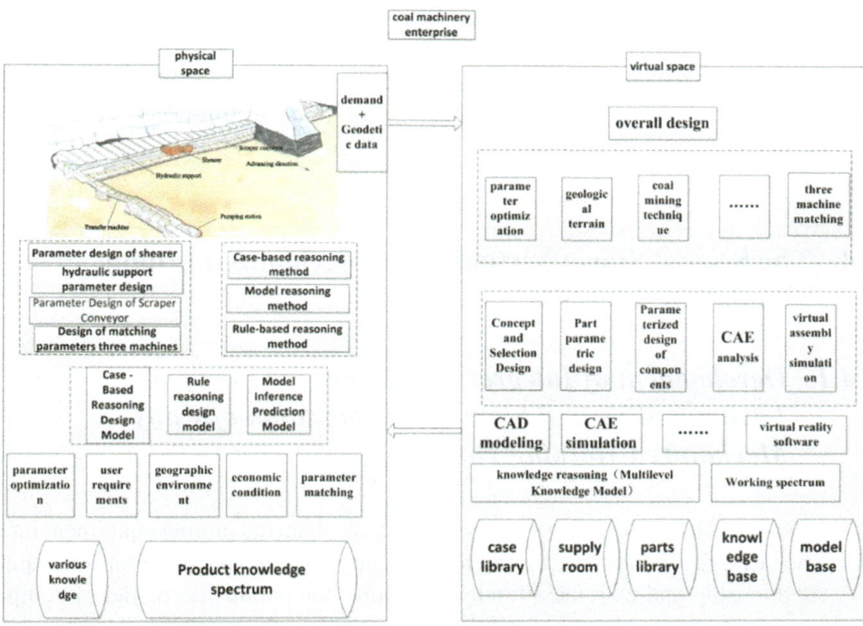

Fig. 2.3 Digital twin Design Diagram for the First Stage of Prototype System

equipment life cycle design process from the early concept selection design stage of fully mechanized mining equipment to the establishment of product digital prototype under real conditions.

After the design of each equipment is completed, the relevant supporting budget work is completed by using the established digital model of fully mechanized coal mining and the complete set of experimental system of fully mechanized coal mining. The equipment virtual supporting is carried out in Unity3D software, and the relevant coal mining process parameters are set for simulation. The specific process is as follows: according to the requirements of mining geographical environment and equipment conditions, the specific model of equipment is selected in the database of fully mechanized mining equipment selection by using the module of equipment selection design and method, and then input into the digital equipment module, and then associate the characteristic parameters in the equipment characteristic size database and equipment performance parameter database, and transfer the parameters into the equipment characteristic simulation script library and Parametric model library of equipment feature parts, then the parametric model of equipment feature parts is generated, and the whole machine assembly model is generated through the whole machine assembly model library, which can fully simulate the real behavior of the equipment. The establishment of the model needs lightweight design, remove the complex internal transmission structure, and be consistent with the actual equipment model in appearance, so as to minimize the cost of server software and hardware resources in virtual simulation. The key parts and the whole machine model can realize data parametric drive, and can fully express the equipment matching operation relationship. Then, the virtual coal seam model is built in Unity3D software, all of which are driven by real-time geological data, and the feature data of coal seam surface can be extracted, which can be transformed into the cutting path that the equipment can run, so as to complete the configuration of comprehensive working face elements.

2.4 The Second Stage: Virtual Fully Mechanized Mining Simulation Analysis

2.4.1 Development of Integrated Software for Element Layout and Virtual Simulation Operation of Fully Mechanized Working Face

(1) The first is how to quickly arrange the fully mechanized mining equipment into the virtual coal seam. According to the equipment matching, motion constraint relationship and extreme attitude conditions, the parametric model of equipment workspace should be established, and then the equipment positioning and operation methods in virtual environment should be studied, including the virtual method of joint positioning and attitude determination of shearer

and scraper conveyor, the coupling analysis method of scraper conveyor and virtual floor, and the positioning and attitude determination method of hydraulic support based on roof and floor curve and geographical environment, virtual method of floating connection between hydraulic support and middle trough of scraper conveyor etc. The above methods are integrated into the design, in order to achieve the purpose of equipment collaborative propulsion simulation in virtual environment.

(2) Research on operation law model of fully mechanized mining equipment. The model is a kind of behavior model which describes the order, randomness, concurrency, linkage and other characteristics of equipment behavior under the action of various possible driving and disturbance. Among them, driving mainly refers to the production plan and planning made in advance; disturbance includes the problems caused by inaccurate geological information detection, or the failure of airborne sensor detection, mainly including the error rate of shearer automatic height adjustment, the problem of drum cutting rock; the problem of scraper conveyor overload; the problem of hydraulic support posture (dumping, mine pressure, roof crushing, etc.)The straightness problem is mainly "three straight and one flat", straight coal wall, straight hydraulic support, straight scraper conveyor, equal cutting top and bottom. The rules include evaluation, optimization, prediction and other rules based on the operation and evolution law of fully mechanized working face.

(3) The main methods of simulation are hardware in the loop simulation and off-line simulation. It can provide the operation support environment of distributed simulation system; it can realize the three-dimensional visual environment simulation demonstration of fully mechanized mining equipment in the geographical environment; the off-line simulation of fully mechanized mining face production system includes fully mechanized mining equipment cutting path independent planning, coal wall model real-time correction; virtual behavior compilation, dynamic operation matching, dynamic parameters real-time variable dynamics, finite element analysis, timely collaboration and integration with engineering analysis software, and whether the virtual state monitoring parameters of fully mechanized mining equipment such as materials and fatigúe are comprehensive, and whether the simulation data can be recorded in real time.

2.4.2 New Method of Simulation Credibility Evaluation

After the simulation method is solved, it is necessary to evaluate the credibility of the simulation results of the simulation system. A new method of intelligent credibility evaluation of composite model under the new fusion VR architecture is adopted to support the credibility evaluation of complex simulation system with the characteristics of "dynamic evolution, reusability and scalability".

2.4.3 *Collaborative Intelligent Evolution Method for Modeling and Simulation of Unmanned Fully Mechanized Mining Equipment System*

After obtaining the simulation results with enough trust, the equipment can be operated as a "robot" in the simulation system first, and an all-round autonomous operation simulation unmanned fully mechanized mining face can be established. On this basis, the modeling and simulation methods such as dynamic complex environment perception, collaborative control, communication interaction and self-organizing collaborative planning method of unmanned fully mechanized mining equipment cluster in virtual environment are studied, which provides a theoretical method basis for the design and evaluation of unmanned fully mechanized mining equipment system.

2.4.4 *Simulation Technology of Real Coal Seam Operation Based on Digital Twin*

Breaking through the key technologies such as virtual real mapping rapid modeling based on real coal seam operation simulation, simulation prediction and process optimization based on twin data simulation, online compensation and precise control of simulation process, the intelligent fully mechanized mining simulation optimization prototype system is developed to realize real-time synchronous operation and precise intelligent closed-loop control of physical mining process and virtual mining process to the purpose of safe and efficient mining.

2.4.5 *Laboratory Cases*

In the second stage, the design schematic diagram is shown in Fig. 2.4. The digital equipment model established in the first stage is imported into Unity3D software, and the digital equipment is arranged in the virtual coal seam according to the actual underground layout rules. By using a variety of key mining parameter models embedded in the bottom layer and the data recorded in real-time operation, different planning conditions are input according to the scheme design for virtual simulation, and the planning data of key parameters are recorded in real time such as the shearer traction speed and the load of scraper convey and hydraulic support follow-up. In the post-processing module, the performance parameters and process parameters are optimized by comparing the virtual simulation results of different schemes, and then run according to the optimization results, so as to achieve the purpose of guiding production. It solves the problems of rapid selection of supporting equipment for coal production enterprises and advance test of pre-selected equipment scheme according to specific underground geological and topographical conditions, finds out various

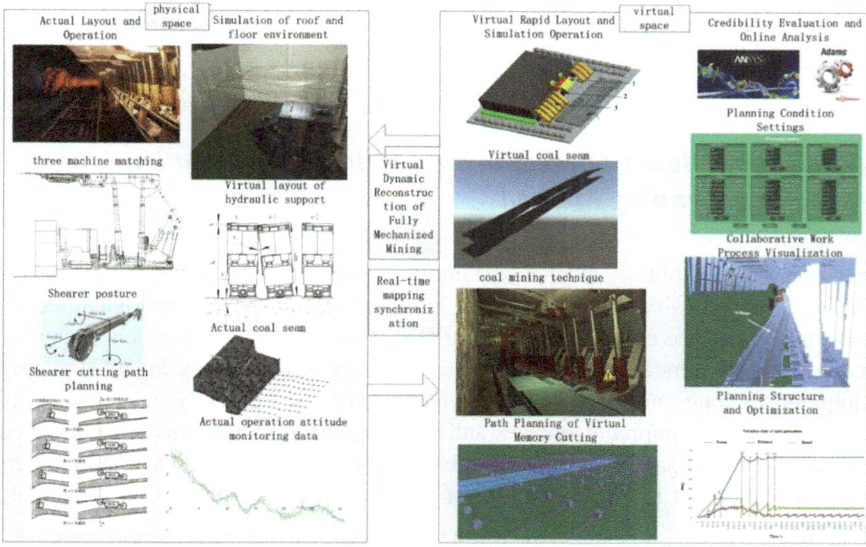

Fig. 2.4 Digital twin design diagram for the second phase of prototype system

problems that may appear in equipment operation in advance, selects the best among many schemes, and achieves the diversity design and optimal design of full life cycle of fully mechanized virtual simulation.

2.4.6 Technical Conception and Practice of "Coal Seam + Equipment" Joint Virtual Simulation Operation in Fully Mechanized Mining Face

In the virtual environment, how to build a virtual production system of equipment self-organizing operation is an important goal of this study, and it can also provide an effective analysis tool for the study of the spatiotemporal movement relationship of equipment operation.Unity3D, a virtual reality software, has the characteristics of easy development, complex scenes, excellent rendering effect, excellent display function, outstanding data interaction function, good multi-platform expansion performance and good data compatibility. In addition to its powerful physical engine function, this paper chooses it as a development tool [14].

In the process of running in fully mechanized working face, the posture of roof and floor after shearer cutting has an impact on the support posture of hydraulic support and the posture of scraper conveyor to a certain extent. In the virtual environment, it is necessary to carry out research from the virtual equipment and coal seam model establishment method, equipment operation simulation method and other aspects, in order to realize the "Coal Seam + Equipment" joint virtual simulation of fully mechanized working face.

2.4.7 Agent Based Three Machine Virtual Collaboration and Planning Method

In the design stage of fully mechanized mining equipment selection, it is often necessary to comprehensively consider many limiting factors, such as geological terrain and coal occurrence conditions, working face parameters and equipment matching and layout requirements. Therefore, the selection cycle of fully mechanized mining equipment is long and difficult. Coal mining enterprises urgently need to quickly select supporting equipment, preview and plan the working conditions of pre-selected equipment scheme, and find out all kinds of problems that may appear in the operation of equipment in advance, the best scheme should be selected among many schemes.

Therefore, all fully mechanized mining equipment elements are digitized, and then all links in the production process are simulated and analyzed at the digital level, so as to predict what will happen after the actual equipment is put into production, which can prevent the losses caused by many problems in the actual production. In the design and selection stage, we can see the whole production process, plan operation details and strategies, improve efficiency, predict possible problems, optimize the whole system, test everything as far as possible at the beginning, simulate and verify the whole production process in the virtual environment, integrate the selection design and process planning, conduct online evaluation and digital planning.

In this section, a VR collaborative planning method based on MAS (Mulit-Agent-System) is proposed, and a prototype system (FMUnitySim) is established, which can perform panoramic three-dimensional visualization rapid planning rehearsal on the working face environment information, type selection design, equipment matching, equipment automation level, equipment performance index, equipment task model and task coordination, coal mining process design integration. The operation state of the whole working face is reproduced completely and truly, which helps coal enterprises optimize, simulate and test in the virtual environment before the actual production, and optimize the whole production process synchronously in the operation and production process of fully mechanized coal mining, so as to achieve safe, efficient and reliable production.

2.5 The Third Stage: Online Virtual Monitoring and Control of Fully Mechanized Production System

2.5.1 *Virtual Monitoring*

- Operation monitoring: collect online data of fully mechanized mining equipment in real time, drive the virtual scene to synchronize, which is used to intuitively monitor the operation status of the whole working face;
- Collaborative monitoring: in the background of monitoring software, the semi supervised theoretical model is used to establish the mathematical model of collaborative operation and carry out real-time calculation, and the virtual monitoring based on data-driven and model fusion is carried out to synchronously predict the operation status of equipment; it is used to improve the reliability and accuracy of monitoring, and has self-correction function;
- Distributed network collaborative monitoring: by using LAN collaborative technology, the real-time running status data of various equipment can be distributed processed, and the monitoring hosts can share and synchronize the status in real time, so as to reduce the network pressure and speed up the data processing speed;
- Dynamic monitoring: secondary development of virtual reality monitoring software, find the engineering analysis software interface, real-time interaction with MATLAB, Adams, ANSYS, AMESim and other engineering analysis software, obtain action data and status. For example: shearer real-time stress state through cutting resistance, cutting motor temperature, current data back to the monitoring host, and through the interface into Adams for dynamic analysis, and then carry out the shearer rocker arm and other key components of ANSYS stress analysis, complete the shearer Adams dynamic analysis, shearer rocker arm and other key components of ANSYS stress analysis and shearer hydraulic adjustment AMESim analysis of the system, and the real-time running analysis results are returned to the virtual monitoring software for real-time presentation.

2.5.2 *Virtual Control*

- Based on the real-time data of the bottom layer, the digital twin simulation platform for the process of fully mechanized mining production system driven by big data is constructed to realize the functions of online real-time virtual operation of intelligent fully mechanized mining face, offline and online simulation and optimization of intelligent fully mechanized mining, so as to form a virtual reconstruction solution for intelligent fully mechanized mining.
- Virtual reverse control: input the real-time optimization control results of intelligent fully mechanized mining virtual simulation into the actual fully mechanized mining equipment control system to realize the optimization of physical system operation control. The verification mechanism of synchronous operation of digital

model and physical working face control system based on data-driven is established to realize the virtual verification and synchronous operation of intelligent fully mechanized coal mining.

- Fully mechanized mining robot completes mining task independently, which is called "unmanned mining". The intelligent fully mechanized mining equipment is upgraded to a multi-agent system for fully mechanized mining. In the complex unknown geological environment or partially known geological environment, which is lack of enough cutting information accuracy, the multi-agent cooperative sensing, planning and control links are integrated. Based on the multi-agent task modeling and allocation model and the information exchange between agents in the case of global normal / local failure/complete failure. Based on the mechanism of mutual and sharing, dynamic self-organization theory and method are used to realize the Autonomous Cooperation of multi intelligent system of fully mechanized coal mining in complex environment, and realize the cooperative cluster motion control of multi robot system of fully mechanized coal mining.

2.5.3 Monitoring System of Three Machines in Fully Mechanized Mining Face Working Condition in Fully Mechanized Working Face Under VR Environment

This chapter first studies and analyzes the "three machine" digital twins theory. Establish "three machine" digital information model, including "three machine" digital model, "three machine" information model and "virtual" and "real" interface, as shown in Fig. 2.5.

Among them, the "three machine" information model is a digital prototype based on the design theory and method of the "three machine" itself and the digital expression of real physical products. The attitude of the "three machine" under the actual

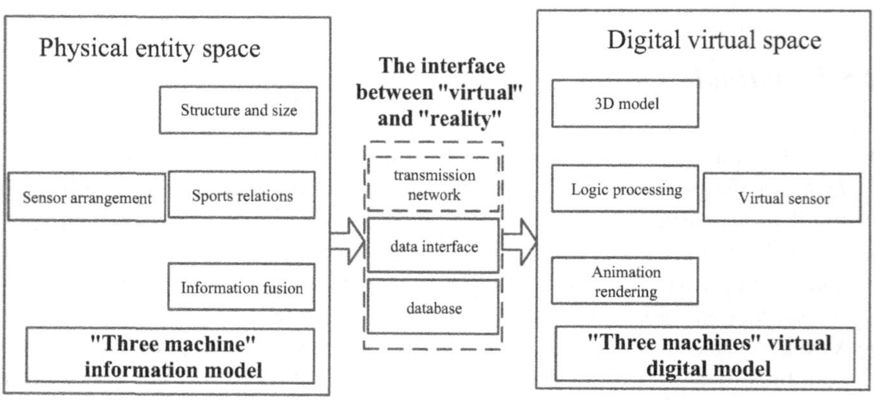

Fig. 2.5 Theoretical system of three-machine Digital Twin

working condition is analyzed, and the synthesis method of synchronous characteristic variables of the "three machine" state is obtained. The information fusion algorithm is studied, and the "three machine" reliability monitoring theory and method are obtained Law.

The virtual digital model of three machines in fully mechanized mining face is mainly a virtual image which is completely consistent with the actual state. In the virtual environment, each component and its surrounding components are constrained and defined, and the interface variables are reserved, which can receive the running state data sent back by three machines in fully mechanized mining face in physical space in real time.

The interface of "virtual" and "reality" can combine the information model and virtual digital model. Through the high-speed network communication platform, it can receive the data sent back by the sensors arranged on the three machines in fully mechanized mining face in real time, transmit it to the database, and transmit it to the centralized control center, remote dispatching room and VR monitoring center in VR laboratory according to the system design requirements Information model interface of test host.

In this way, the operation status of the three machines in fully mechanized mining face can be accurately simulated and real-time monitored and synchronized. The operator can interact with the system, enter any area of the system simulation at any time through any space, observe the operation condition of the equipment, give an alarm to the abnormal situation, and timely find and deal with the faults and problems existing in the operation.

2.5.3.1 Overall Framework Design of "Three Machine" Condition Monitoring System

(1) System design objectives

This monitoring system is based on VR technology, through the data sent back by the sensors of the Three machines in fully mechanized mining face equipment in the fully mechanized working face in real time, the corresponding VR monitoring picture is driven in real time, and the virtual image of the "three machines" in the fully mechanized working face in real time is established, so that the real-time running state of the working face can be displayed in a three-dimensional panoramic view with low delay, clear picture and no jamming. According to the demand analysis of the system, the main design objectives of the system are as follows:

- The VR model and environment which are completely consistent with the real Three machines in fully mechanized mining face are established, and the corresponding control scripts are written, so that the actual Three machines in fully mechanized mining face running state can be fully displayed in the virtual screen;

- The virtual "three machine" model needs to reserve the interface of each sensor characteristic variable, which can easily import the real-time face equipment data that has been stored in the database;
- Sensors that can reflect the pose changes of Three machines in fully mechanized mining face should be arranged on the three machines in fully mechanized mining face in the actual working face, and can be reasonably arranged on each equipment, without affecting the normal work of the equipment;
- In the process of monitoring, the data should be stored, and the historical data should be fully analyzed and modeled from the perspective of big data, so as to predict the operation status of the whole fully mechanized mining equipment;
- In the monitoring mode, due to the large number of equipment in the fully mechanized coal mining face and the large amount of sensing information, if only one monitoring host is owned, it will inevitably cause huge pressure on the server, the operation of VR monitoring screen is not smooth or the presence of stuck scene, which seriously affects the reliability of VR monitoring. Therefore, it is urgent to study a technical solution to solve this problem.

(2) Overall design
According to the above system design objectives, the technology, process route and research methods of the system are as follows:

- UG is used to model the actual parameters of "three machines" in fully mechanized coal mining face. The model is repaired in 3DMAX, and then the script is written in Unity3D software. The panoramic VR scene of fully mechanized coal mining is established, and the running state variables of shearer, scraper conveyor and hydraulic support are set;
- Read and update the real-time operation status data of each equipment in SQL Server database, drive the corresponding action of each equipment, accurately depict the status of underground roof and floor through various sensors, and store the data in the working face database in real time;
- The bottom layer uses the DLL file written by MATLAB software as the data monitoring and analysis module to alarm the abnormal situation and find and deal with the faults and problems existing in the operation in time;
- In view of the large number of equipment in fully mechanized mining face, the server pressure caused by single host monitoring is large, and the VR monitoring screen is not running smoothly or stuck. Using the method of LAN cooperation, multiple hosts synchronize data with each other at any time through the C/S architecture, and then synchronize the pictures in real time to synthesize a panoramic picture of the whole working face;
- The monitoring mode of the system adopts two-way mode: the VR monitoring system in this chapter is integrated on the centralized control console of the central control center of the channel, and is integrated with the video monitoring, data monitoring and the central control center to complement

each other. In the future, a display can be installed in the central control center for VR monitoring.

2.5.3.2 VR Monitoring Method Based on Unity3D

VR monitoring scene can display the operation status of the whole fully mechanized working face in 3D form in real time. It mainly depends on the data sent back by the working face in real time. Therefore, it is necessary to establish the virtual model and virtual scene which are completely consistent with the actual fully mechanized mining scene. Only when the algorithm and formula are programmed and the interface is reserved to read the real-time data of working face, can it keep synchronization with the equipment state of working face.

- Reservation of state variables in VR environment
- Method of real time reading and accessing data
- Real time calculation method of underlying mathematical model
- Real time rendering of coal mining environment information
- Display of fault occurrence screen
- Realization of real-time switching video surveillance screen in VR environment
- Virtual monitoring and real time synchronization method based on LAN.

2.5.3.3 Multi Software Real Time Coupling Strategy

The network I/O module connects the sensor information to the high-speed network communication platform, and accesses the Kingview monitoring host through Modbus TCP protocol. The Kingview monitoring system can upload the collected data to SQL Server2008 database in real time, and the VR monitoring host can call the data in the same local area network database in real time to complete the connection of the data to the VR host. Then Unity3D, MATLAB and SQL Server software interact in real time.

- Kingview + SQL Server
- SQL SERVER + Unity3D
- Matlab calculation result processing.

2.5.3.4 Prototype System Development

After the system design and key technology research, the prototype system is developed. The design of the monitoring panel and the setting of the relevant operation buttons, the design of the relevant monitoring perspective, according to the perspective of the camera arranged in the real working face for rapid monitoring, the main picture shows the panorama, the auxiliary picture and the main picture complement each other to complete the real-time monitoring of the operation state of the fully mechanized coal mining.

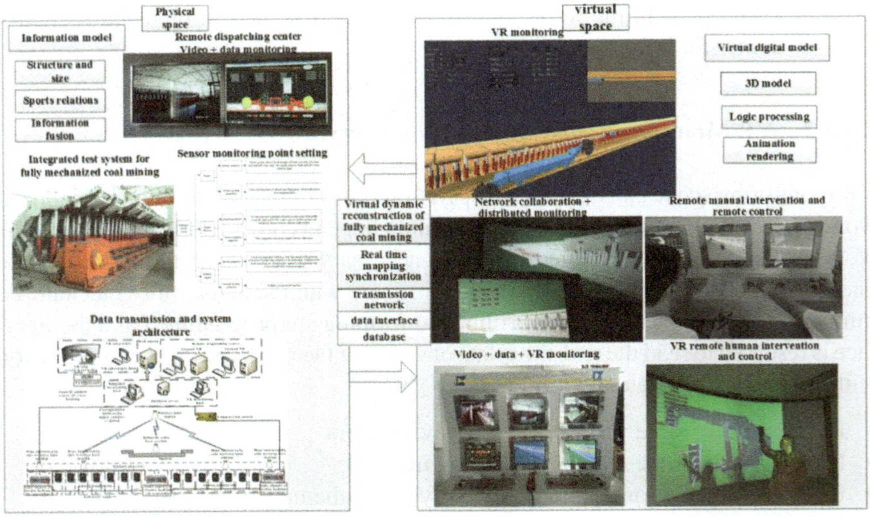

Fig. 2.6 Digital Twin Design Diagram for the Third Phase of Prototype System

VR monitoring can be compared with the real video monitoring system of underground fully mechanized working face, and complement each other to monitor the operation status of fully mechanized working face in real time.

2.5.4 Cases

The design schematic diagram of the third stage of fully mechanized virtual monitoring is shown in Fig. 2.6.

The laboratory has a complete set of coal mine fully mechanized mining test system, has completed the intelligent transformation, installed a centralized control center, equipped with hydraulic support electro-hydraulic control system, working face intelligent control system and video monitoring system, has the ability to simulate the actual underground fully mechanized mining equipment kinematics, and can realize remote automatic coal mining. Relevant sensors are arranged on the whole set of equipment and the artificial roof and floor environment to collect data in real time and transmit them back to the centralized control center, remote dispatching center and VR laboratory through high-speed communication network.

The compiled equipment virtual reality system is integrated and summarized to complete the virtual reality system of fully mechanized mining equipment; the variables of equipment digital model are extracted, the data interface is established, and the real-time sensor information is accessed for real-time data acquisition; the real-time data drives the virtual fully mechanized mining face, and VR monitoring can be carried out on the three platforms respectively. The operator can use the "virtual picture + Video + data" to press the button on the operation panel of the simulated central control center to carry out remote manual intervention on the operation of the equipment.

It can also use a variety of virtual reality human–computer interaction means and virtual human–computer interaction interface to carry out virtual operation in VR laboratory. Virtual reality hardware and software can capture the position, posture and action of the operator, convert the virtual operation into real instructions and access the centralized control center to operate the real equipment, guide the corresponding equipment to complete the adjustment work, and then complete the operation To achieve the task of remote manual intervention and inspection, and to present the real-time operation condition of fully mechanized working face, and then to monitor. In the safe place far away from the production site, remote manual intervention is directly carried out on the abnormal equipment, which supports multi person cooperation and simultaneous inspection.

2.6 Digital Twin Development Trend of Production System in Fully Mechanized Working Face

At present, most of the researches on virtual reality in the field of coal mine are partial or simple. There are not many researches that integrate the achievements of various stages and fields to form a global research. At present, the breakthrough overall research achievements have not yet appeared. Therefore, how to integrate local research results and form a systematic global application platform is very urgent. In this paper, it is suggested that a national strategic plan for the research and application of virtual reality technology in the coal industry should be formulated as soon as possible and implemented in four steps.

The first is to strengthen the integration research of basic theories in various fields, especially the research on the correlation, intersection and integration between the telepresence of technical training, the reality of data visualization and virtual planning.

The second is to strengthen the research of virtual monitoring in the whole production process, solve the problems of virtual monitoring in various fields and links, and provide basic support for the latter two steps.

The third is to implement the research of coal machine whole life virtual monitoring manufacturing, solve the virtual design of coal machine equipment manufacturing and form a mature intelligent equipment manufacturing system.

The fourth is to realize the research of transparent unmanned mining, build a complete and comprehensive intelligent platform which is easy to control between the underground and the well, between the virtual monitoring and real production, between the overall control and local links of the mine, and then build a perfect transparent unmanned high intelligent system which integrates mining, transportation, lifting and selection.

References

1. Mohammad, M. N., Sulaiman, N., Muhsin, O.A.: A novel intrusion detection system by using intelligent data mining in weka environment. Procedia Comput Sci. **3**(1), 1237–1242 (2011)
2. Wang, X., Yang, Z., Ding, H., et al.: Research and application on coal mine machinery equipment cloud manufacturing resource service platform. J. China Coal Soc. **38**(10), 1888–1893 (2013). (in Chinese)
3. Ding, H., Yang, Z.: Method and system of shearer cutting unit modern design oriented to KBE. J. China Coal Soc. **37**(10), 1765–1770 (2012). (in Chinese)
4. Yang, Z., Wang, X.: Study on cloud simulation CAE service system for mining machinery equipment. J. Mech. Eng. **49**(19), 111–118 (2013). (in Chinese)
5. Xie, J., Yang, Z., Wang, X., et al.: A virtual reality collaborative planning simulator and its method for three machines in a fully mechanized coal mining face. Arab. J. Sci. Eng. **43**(9), 4835–4854 (2018)
6. Xie, J., Yang, Z., Wang, X., et al.: A joint positioning and attitude solving method for shearer and scraper conveyor under complex conditions. Math. Probl. Eng. **2017**, 1–14 (2017)
7. Sun, Y., Hua, G., Dou, L., et al.: Study and challenges on cyber-physical system of mine engineering. Coal Sci. Technol. **46**(2), 103–111 (2018). (in Chinese)
8. Tao, F., Liu, W., Liu, J., et al.: Digital twin and its potential application exploration. Comput. Integr. Manuf. Syst. **24**(1), 1–18 (2018). (in Chinese)
9. Tao, F., Cheng, Y., Cheng, J., et al.: Theories and technologies for cyber-physical fusion in digital twin shop-floor. Comput. Integra. Manuf. Sys. **23**(8), 1603–1611 (2017). (in Chinese)
10. Tao, F., Zhang, M., Cheng, J., et al.: Digital twin workshop: a new paradigm for future workshop. Comput. Integra. Manuf. Syst. **23**(1), 1–9 (2017). (in Chinese)
11. Zhuang, C., Liu, J., Xiong, H., et al.: Connotation, architecture and trends of product digital twin. Comput. Integra. Manuf. Syst. **23**(4), 753–768 (2017). (in Chinese)
12. Luo, W., Hu, T., Zhang, C., et al.: Digital twin for CNC machine tool: modeling and using strategy. J. Ambient. Intell. Humaniz. Comput. **10**(3), 1129–1140 (2019)
13. Guo, J., Zhao, N., Sun, L., et al.: Modular based flexible digital twin for factory design. J. Ambient. Intell. Humaniz. Comput. **10**(3), 1189–1200 (2019)
14. Zheng, Y., Yang, S., Cheng, H.: An application framework of digital twin and its case study. J. Ambient. Intell. Humaniz. Comput. **10**(3), 1141–1153 (2019)
15. Zhang, H., Liu, Q., Chen, X., et al.: A digitaltwin-based approach for designing and multi-objective optimization of hollowglass production line. IEEE Access. **5**(2017), 26901–26911 (2017)
16. Tao, F., Zhang, M.: Digital twin shop-floor: a new shop-floor paradigm towards smart manufacturing. IEEE Access. **5**, 20418–20427 (2017)
17. Patterson, E.A., Taylor, R.J., Bankhead, M.: A framework for an integrated nuclear digital environment. Prog. Nucl. Energy **87**, 97–103 (2016)
18. Heidari Jozam, M.M., Allameh, E.E., Vries, D.B.B., et al.: Smart-BIM virtual prototype implementation. Automat. Constr. **39**(4), 134–144 (2014)
19. Tang, T., Teng, L., Wu, J., et al.: Fully realizing digitization is the only way to intelligent manufacturing. China Mech. Eng. **29**(3), 366–377 (2018). (in Chinese)
20. Tao, S., Sui, F., Liu, A., et al.: Digital twin-driven product design framework. Int. J. Prod. Res. **1**, 1–19 (2018). https://doi.org/10.1080/00207543.2018.1443229

Chapter 3
Virtual Simulation Element Model Transformation Technology and Model Base

3.1 Introduction

Coal machinery equipment virtual reality model resource library is the basis of the establishment of coal machinery equipment oriented virtual reality assembly system. According to the different characteristics of shearer, scraper conveyor, road header and hoist structure and actual work, the structure is divided in detail, and the virtual reality model resource library is established. In the follow-up work, it is further realized by programming to complete the designed functions.

3.2 Framework Design

The virtual reality resource library is a series of OSG or Ive files arranged according to a certain classification, so that the corresponding functions can be realized by programming in the following chapters, which are divided into virtual disassembly resource library and scene simulation resource library, as shown in Fig. 3.1.

- The virtual disassembly resource library is divided into two parts. The first part is the model of a single part in a scene and the whole scene of automatic assembly and disassembly, which provides conditions for making corresponding operation scene and automatic demonstration scene respectively;
- Scene simulation resource library is a single made automatic demonstration scene file, which completes the corresponding functions through programming.

Virtual reality resource library production technology mainly includes: structure level division technology, CAD modeling technology, CAD model conversion technology, CAD model restoration technology, Google 3dwarehouse resource search and download technology and 3DMAX scene and animation production technology. These technologies are linked together and constitute the main content of virtual reality resource library production technology, as shown in Fig. 3.2.

X. Wang et al., *Virtual Reality Technology in Mining Machinery*,
https://doi.org/10.1007/978-981-16-4408-5_3

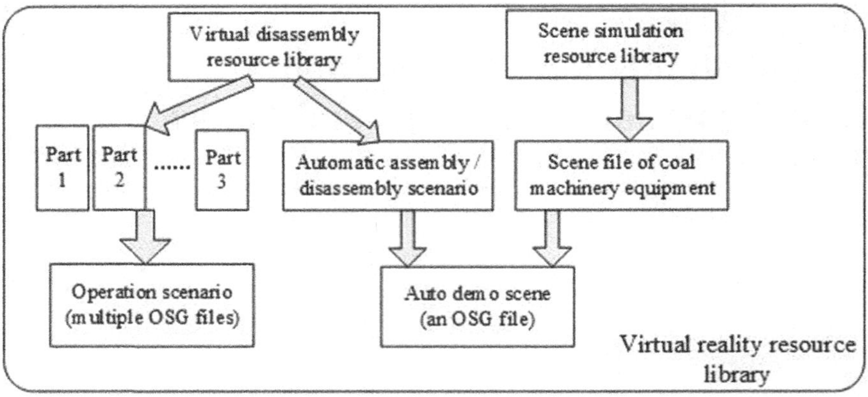

Fig. 3.1 Framework of virtual reality resource library

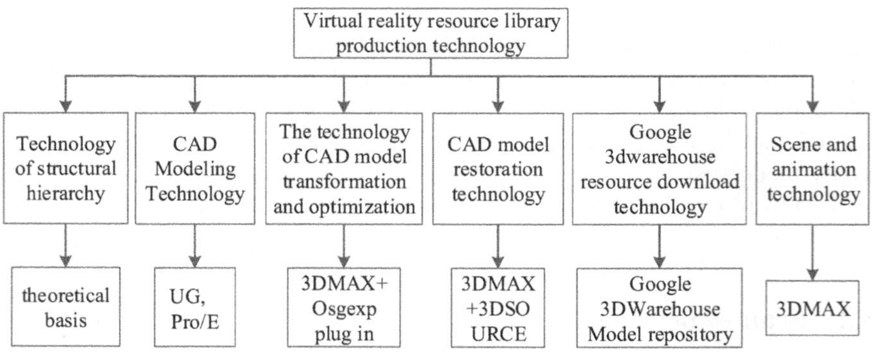

Fig. 3.2 Virtual reality resource library production technology

3.3 Technology of Structural Hierarchy

There are many parts and components of coal machinery equipment, and the structure is complex. Therefore, in order to divide the equipment structure of coal machinery equipment, we must have a deep understanding of its corresponding internal structure and work process. We should learn relevant books and materials in the early stage, and then practice in the corresponding production enterprises, so as to lay a good theoretical foundation.

In the first mock exam, we need to consider the factors of function, size, whole and hierarchical assembly. We need to have the relationship between the whole and the part. We will assemble parts that belong to one functional module individually, then assemble the module as a whole and the same level module, then assemble the first level components to the shearer. For example: the shearer can be divided into

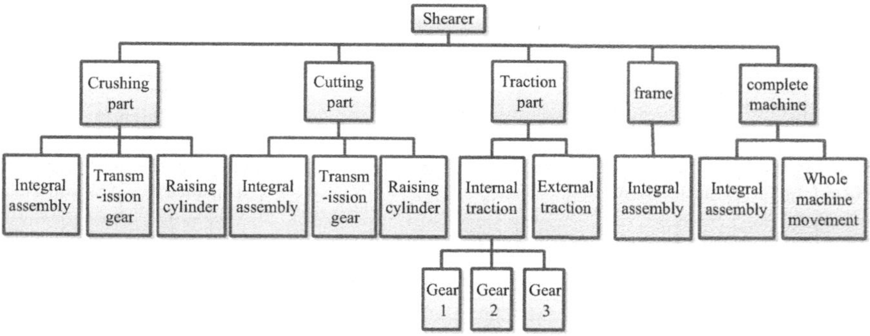

Fig. 3.3 Structure classification diagram of Shearer

crushing part, cutting part, traction part and frame according to its function, which overall classification is shown in Fig. 3.3.

3.4 CAD Modeling Technology

In order to make the coal machine equipment model be realistic, this book takes the coal mining machine modeling as an example to analyze. The modeling of the model is divided into three steps: the first step is geometric modeling, which mainly establishes the geometry of the assembly model; the second step is physical modeling, which mainly deals with the color, material map and lighting of the geometric model; the third step is behavior modeling, which mainly deals with the motion and behavior description of virtual model.

The geometric modeling tools are generally divided into art and engineering. The models established by 3DMAX and other art software are not generally accurate, which is not conducive to the extraction of virtual prototype data information. Such software is not suitable for modeling such large mechanical equipment as shearer, while engineering software such as UG cannot be directly converted into virtual reality model, so it needs to be converted in a certain way. Therefore, in order to make up for the shortcomings of UG and 3DMAX and make full use of their advantages, this book uses 3DMAX as the transition software after establishing the accurate model in CAD modeling software, and then transforms and modifies the model to generate high-quality virtual model.

Both UG and Pro/E are modeled from bottom to top by assembly tree structure. As shown in Fig. 3.4, the model composition diagram of scraper conveyor is shown. The figure not only shows the composition and assembly sequence of actual coal machine equipment, but also shows the relationship between assembly body and assembly unit.

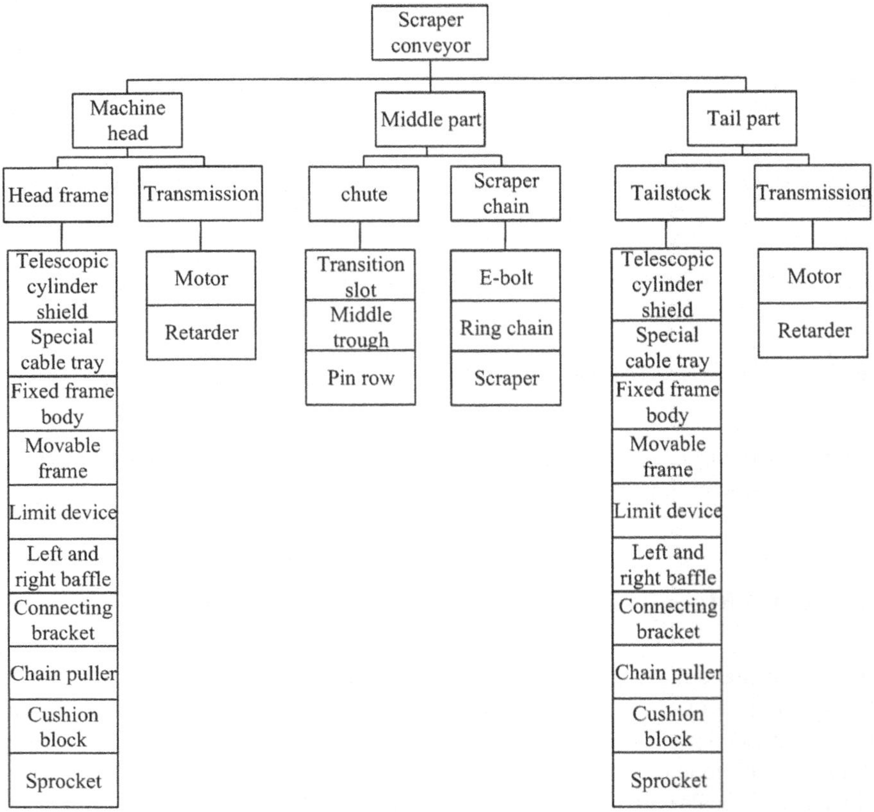

Fig. 3.4 Composition diagram of scraper conveyor model

According to the above model composition diagram, this paper uses UG for shearer and Pro/E for scraper conveyor to carry out supplementary modeling on the basis of the original part of the model, forming the complete model of the two and their mutual cooperation model. The specific CAD model is shown in Figs. 3.5 and 3.6.

The model format established by UG is different from that established by pro/e. In order to facilitate the conversion of the subsequent models, the model that the two cooperate with will be set in the next step 3DMAX.

Fig. 3.5 UG model of shearer

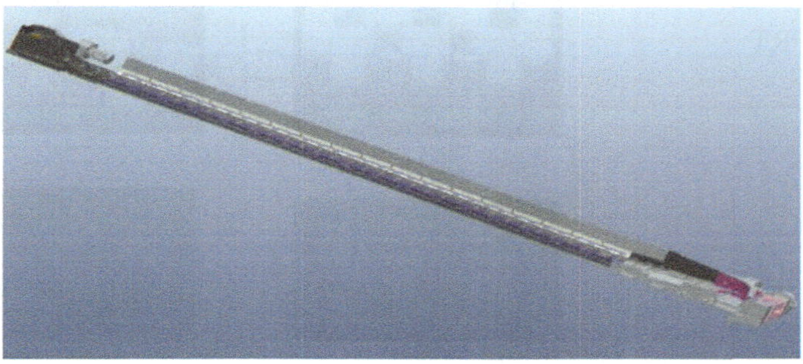

Fig. 3.6 Pro/E model of scraper conveyor

3.5 CAD Model Transformation and Optimization Technology

Taking the track shoe of the driving part of the road header as an example, the conversion process and results of the above four intermediate formats exported in UG into 3DMAX are compared. The results are shown in Fig. 3.7 and Table 3.1.

Similarly, select a gear in the shearer, and compare the conversion process and results of various formats acceptable from UG to 3DMAX, as shown in Fig. 3.8 and Table 3.2.

Similarly, the paper selects the pin row of scraper conveyor to carry out the format conversion experiment from pro/e to 3DMAX, and obtains the following conversion effect Fig. 3.9 and characteristic comparison Table 3.3.

Through the comparison of the first mock exam and the analysis of the conversion characteristics of various formats, we can get that the IGS in the middle format model is transformed from the same model. The model of IGS format and WRL format occupies a large amount of memory, which can indirectly slow down the running speed of the system, while STL is moderate, and DWG is the smallest; in terms of conversion time, the conversion time of IGS format is more, while the conversion

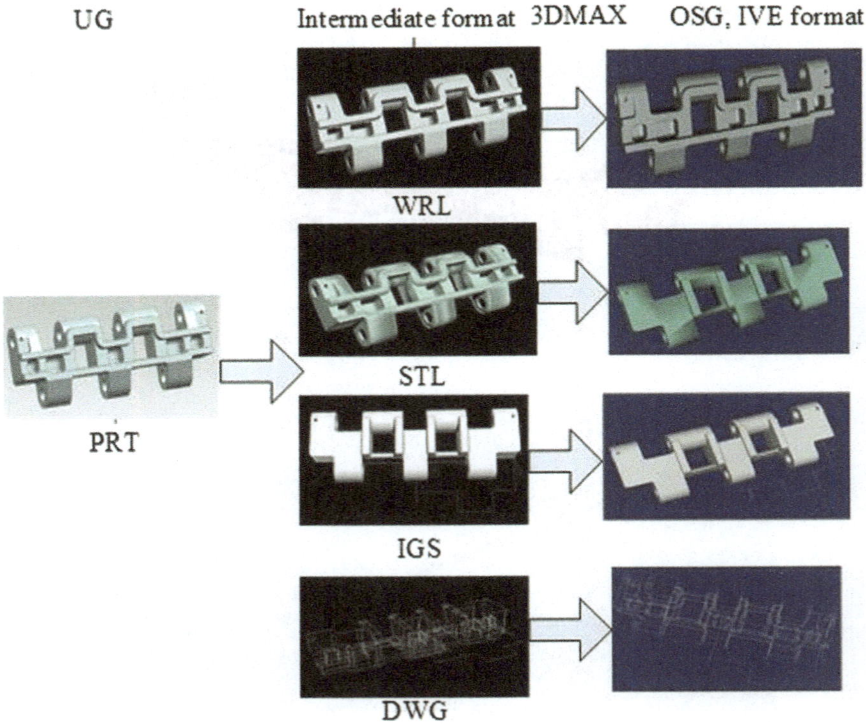

Fig. 3.7 UG track shoe model conversion

Table 3.1 UG conversion 3DMAX and OSG format comparison

Format	Intermediate format model size (KB)	Conversion time (s)	OSG, IVE format size (KB)	Characteristic
WRL	420	2.0	516	A model is divided into multiple editable meshes and cannot be exported to a single model
STL	300	3.4	421	The effect of the model is good, and the material is not supported
IGS	1619	15.6	941	The phenomenon of face breaking is serious
DWG	40	1.2	1961	Line model, no face

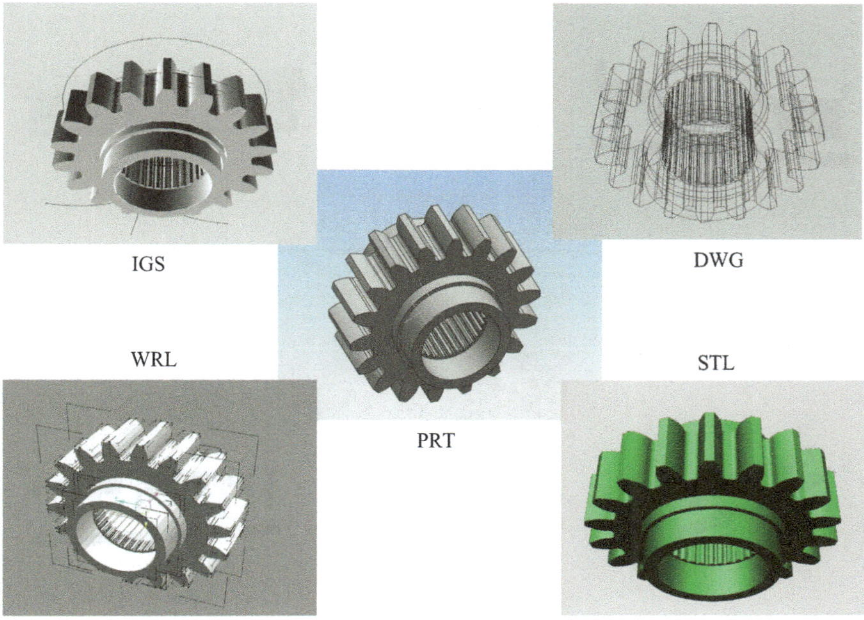

Fig. 3.8 UG gear model conversion

Table 3.2 UG conversion 3DMAX format comparison

	PRT	IGS	DWG	WRL	STL
Size (KB)	2001	3210	103	518	500
Time (s)	0	25.4	1.3	2.1	6.5
Role	Initial format entity model	The surface is damaged or missing	Line model, no face	The surface is not smooth, and a model is divided into multiple editable meshes	The effect of the model is good, and the material is not supported

time of the other three formats is shorter; from the size of converted OSG and IVE format files, the OSG format size of DWG and IGS is significantly larger than that of WRL and s The size of TL format is larger than that of WRL and DWG format. Considering the particularity of coal machinery equipment, it can be concluded that the STL format is more suitable for conversion. The conversion time of STL format is faster, the memory of the model is smaller, and the real sense of the original model can be maintained after conversion, which can meet the basic needs of virtual assembly and scene simulation.

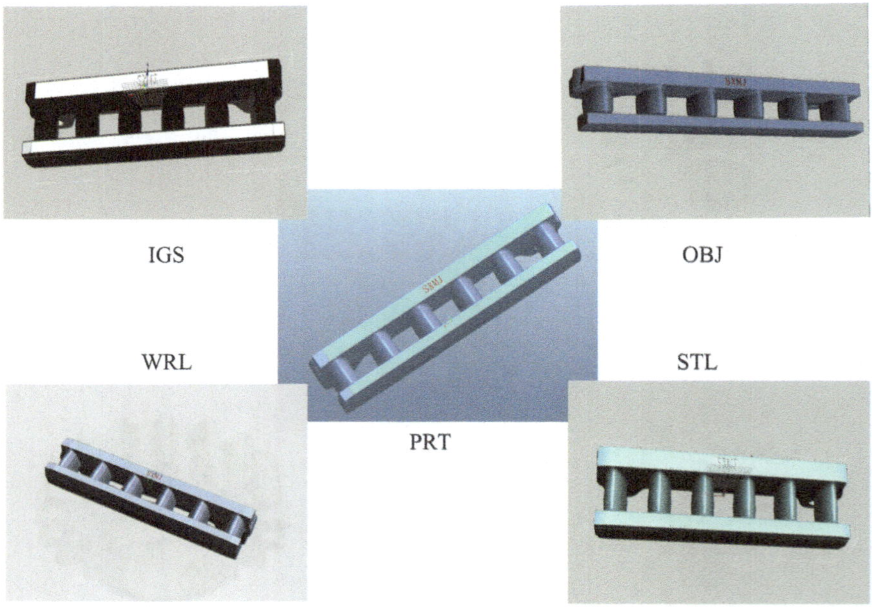

Fig. 3.9 Pro/E pin layout model conversion

Table 3.3 Comparison of 3DMAX format in Pro/E conversion

	PRT	IGS	OBJ	WRL	STL
Size (KB)	3819	5040	1507	592	218
Time (s)	0	36.3	1.5	2.6	1.7
Role	Initial format entity model	The surface is damaged or missing	The model is effective and difficult to operate	The surface is not smooth, and a model is divided into multiple editable meshes	The effect of the model is good, and the material is not supported

After finding out the intermediate format method of model transformation, how to complete the optimization of the transformed model has become a key problem because of the huge model of coal machinery equipment. For virtual assembly and scene simulation, model optimization can be carried out from the number, size and rendering effect of models. The number and size of models affect the reading, operation and operation of the whole virtual assembly system. If there are too many and too large models, it will lead to slow reading progress, unstable system operation, and lagging operation. There is no primary and secondary difference in the whole system. The highest level of virtual reality is the integration of virtual and reality, which cannot be separated from each other. Good rendering effect can ensure the system and the reality of the model. There are three optimization methods.

(1) The combination and separation of models. When transforming through STL, you need to select the exported models. Selecting multiple models to transform together will make these models merge into a whole. It is generally recommended that each model be transformed separately. For the parts that have no assembly task or are integrated, they are generally exported as a whole. However, if the surfaces of two parts are in contact or interference, they should be exported separately, otherwise, it is very easy to flip the model surface method. In addition, because there are many parts of coal machinery equipment and there are primary and secondary parts, the main ones are the more important and key parts that reflect the assembly relationship. These parts have assembly tasks, and they must be converted individually one by one. The secondary ones are the accessory parts of the key parts. They can be merged and converted into one part, which can not only reduce the assembly time After removing the assembly tasks of these accessory parts, the assembly process of important and key parts can be reflected more clearly and in detail. There are four models before transformation, and one model after transformation.
(2) The selection of conversion parameters. After selecting STL as the intermediate conversion format, the selection of intermediate conversion parameters becomes the key. Appropriate conversion parameters can make the model obtain good display effect, and inappropriate parameter settings will produce model effect that does not meet the system design requirements, as shown in Fig. 3.10.

Figure 3.10 shows the effect of the model when the precision of STL parameters is 0.01, and the figure below shows the effect of the model when the precision is set to 40. The upper left corner of the two figures shows the number of faces of the model. When the precision is 0.01, the number of faces is 147,723, and when the precision

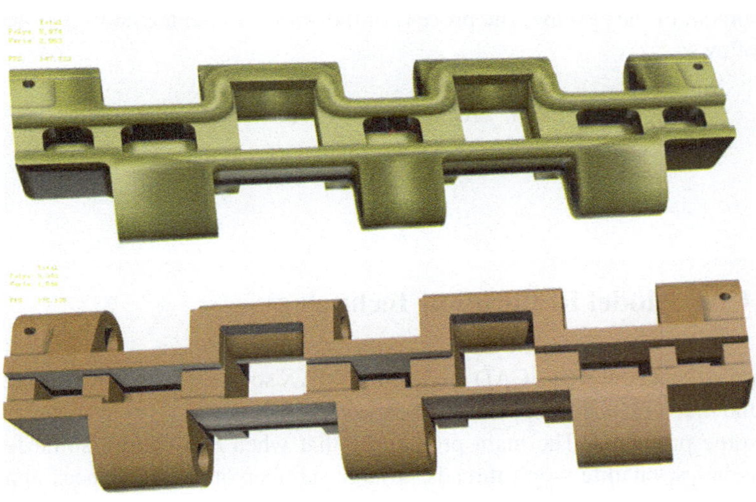

Fig. 3.10 Wrong parameter setting

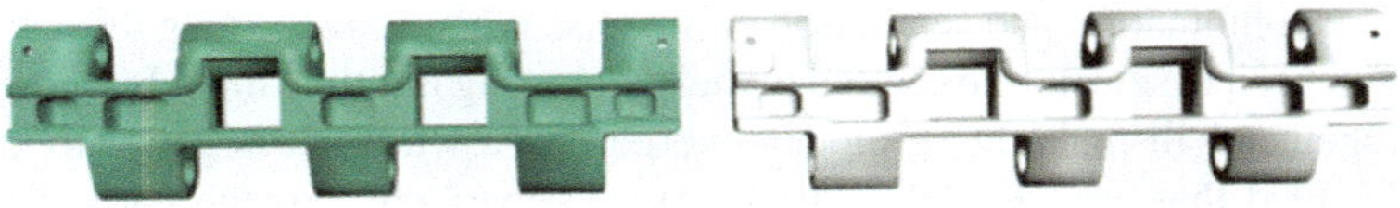

Fig. 3.11 Effect of model before and after rendering

is 20, the number of faces is 17,513. Therefore, the smaller the precision is, the more faces of the model after conversion, the more refined the model is The greater the pressure caused by row, the less the number of model faces and the rougher the model is, the more often the straight edges and right angles appear, and the display effect cannot meet the requirements, so it is necessary to seek the balance between the two.

(3) Rendering of the model. There are two kinds of model rendering: material and map. Because the map needs to introduce external pictures, it will increase the memory occupied by the model, and the material can set the model as the sensory experience of the real material, which is in line with the real effect expected by the virtual reality system. Therefore, the effect of the model after rendering is more realistic than before. In 3DMAX, metal material is given to it, and the parameters are set as follows:

Ambient brightness: 170; diffuse brightness: 150; high light level: 50; gloss: 50; self-luminous color: black; opacity: 100. The comparison of the effect before and after is shown in Fig. 3.11.

Finally, according to the composition diagram of coal machine equipment model, each part of the model is transformed into OSG or Ive format recognized by virtual reality software through osgexp plug-in, which is convenient for the subsequent development of the system. The process of the whole model technology processing is as follows:

$$\text{UG, Pro/E} \rightarrow \text{3DMAX} \rightarrow \text{VR}$$
$$\text{PRT} \rightarrow \text{STL} \rightarrow \text{OSG, IVE}$$

3.6 CAD Model Restoration Technology

In the process of converting CAD model to 3DMAX software in the previous section, we found that STL is the best intermediate conversion format. However, we encountered some problems. The main problem is that when STL shows some features such as holes, chamfers and threads, a large number of polygon faces appear in the model. This is because 3DMAX and other art software generally use polygon mesh algorithm, while UG and other engineering software use NURBS algorithm

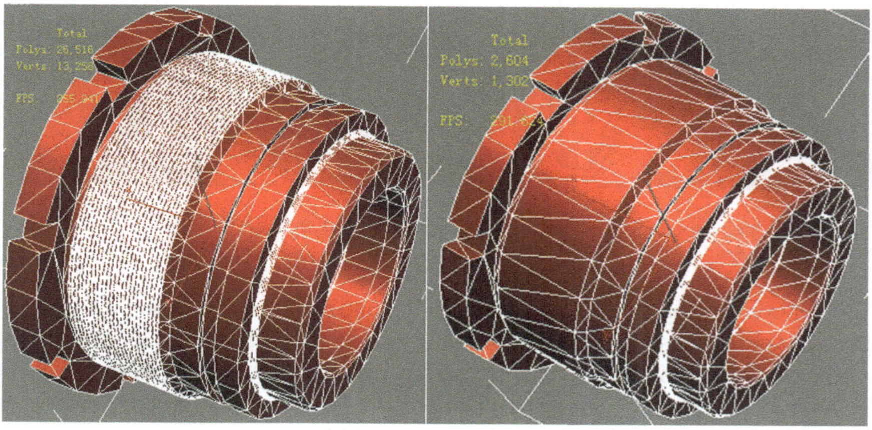

Fig. 3.12 Comparison of guide sleeve number of height adjusting cylinder in crushing section

[1], especially In fact, these are in a relatively secondary position in virtual assembly and scene simulation. Therefore, it is necessary to simplify and repair the threaded holes and chamfers of these parts in UG.

Figure 3.12 is a comparative diagram of the influence of the screw thread on the number of faces in the guide sleeve of the height adjusting cylinder of the crushing section. It can be seen from the figure that the number of faces in the guide sleeve with and without screw thread is nearly ten times different.

The final running speed of VR system is affected by three factors: the total number of VR scene models, the total number of VR scene models, and the total amount of VR scene models. For the virtual assembly system, the first two are the main factors, so in the virtual reality assembly system for coal machinery equipment, the optimization of the model is not only to simplify the number of faces of each independent model, but also to simplify the total number of models.

Too many models will directly affect the export of VR scene and the opening speed of VR scene. In the virtual reality model repository system, for some parts without assembly tasks, merging them into one part not only makes the assembly more convenient, but also realizes the optimization of the model.

For the optimization of the number of faces, manual optimization and software optimization are mainly used. Manual optimization is mainly used to repair some broken surface and broken line after model transformation. It can also optimize other aspects. For software optimization, you can use your own 3DMAX script file or some other surface reduction tools. This chapter uses the software polygon Cruncher to optimize. Polygon Cruncher is a plug-in supporting 3DMAX. Its main function is to minimize the number of polygons in the 3D model without affecting its appearance.

As shown in Fig. 3.13 is the local effect picture of the optimization of the rocker arm of the crushing part of the shearer. The left picture shows the non-optimized model, the middle picture shows the 50% optimized model, and the right picture shows the 80% optimized model.

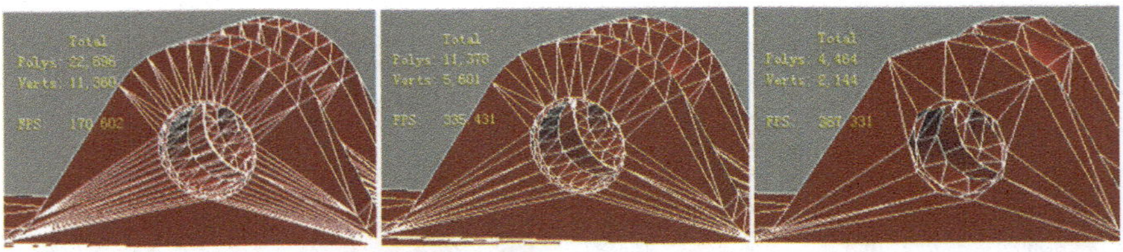

Fig. 3.13 Comparison of local effect of rocker arm optimization in crushing part

After completing the transformation and optimization of coal machinery equipment, it is necessary to render and bake the appearance of coal machinery equipment, so as to make the final appearance of each group of coal machinery components in the large system realistic and enhance the reality of the system. Figure 3.14 shows the rendering effect of the whole Shearer in 3DMAX.

After all operations of coal machine model are completed, it is necessary to export it to OSG recognizable OSG or IVE format through osgexp plug-in, so that OSG can transfer the model into the virtual assembly scene to complete the functions provided by the system[2].

Fig. 3.14 Final rendering effect of shearer model

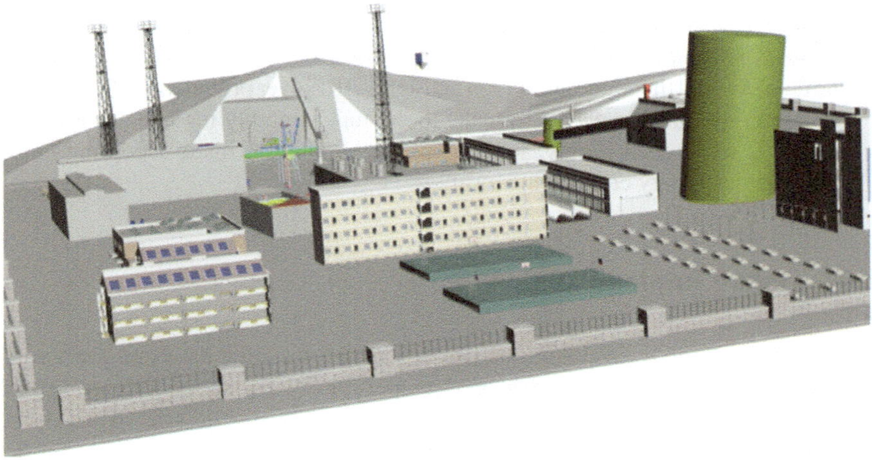

Fig. 3.15 Model of modern industrial square of coal mine

3.7 Search Technology of Google 3dwarehouse Resources

This virtual reality resource library includes many buildings, involving building blocks. Learning the modeling of these buildings requires a lot of time and energy. If we can use the existing online 3D resources to make some scenes, it is very critical. Google 3dwarehouse model library has now become very successful, with a large number of 3D models available for users to download. You can use this model library to download the required 3D materials [3].

In the establishment of modern industrial square model of coal mine, the administrative building, industrial plant, lifting machine room, warehouse, dormitory building and other models are all from Google 3dwarehouse. Some information about the layout of coal mine industrial square is searched, and the model of coal mine industrial square is established [4]. The effect is shown in Fig. 3.15.

3.8 3DMAX Scene and Animation Technology

As a powerful 3D design software, 3DMAX integrates modeling, animation and rendering. At the same time, it also plays an intermediate role in the system.

- File format conversion. To import virtual reality environment from UG, Pro/E and other CAD modeling software needs to be transformed by 3DMAX software.

- Document effect production. Need to set the material in 3DMAX, this system is mostly set with metal texture material, in order to import the virtual reality environment in the future to have a better effect.
- Animation technology. In 3DMAX, the imported model is animated according to the demand, and the frame animation and spline IK solver provided by the software are used. The specific way is to make the virtual assembly module according to the best path of disassembly, and the scene simulation is made according to the real coal machinery equipment workflow.

3.9 System Implementation

According to the above technology of virtual reality resource database, the virtual disassembly resource database of shearer, scraper conveyor, road header, elevator and scene simulation resource database are established respectively. The virtual disassembly resource database is introduced with road header as an example. Virtual assembly of road header mainly includes eight parts: cutting part, shovel board part, transportation part, body part, walking part, rear support part, hydraulic system and whole machine. Figure 3.16 is the scattered and assembly diagram of secondary components of road header made in 3DMAX software.

Figure 3.17 shows the vertical shaft skip hoisting system. Through the car dumper—bottom coal bunker—coal feeder—loading equipment, the coal enters the skip. The skip is lifted from the skip loading chamber to the wellhead coal bunker under the action of wire rope, crown wheel and hoist. In this way, the transportation and hoisting process of coal is completed, which can basically reflect the actual application of the equipment in the whole vertical shaft skip hoisting system The overall and local operation state of the system [5].

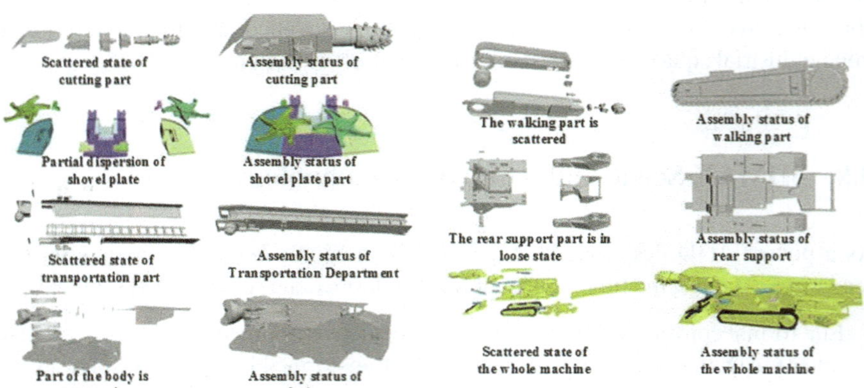

Fig. 3.16 Distribution and assembly drawing of each part of road header

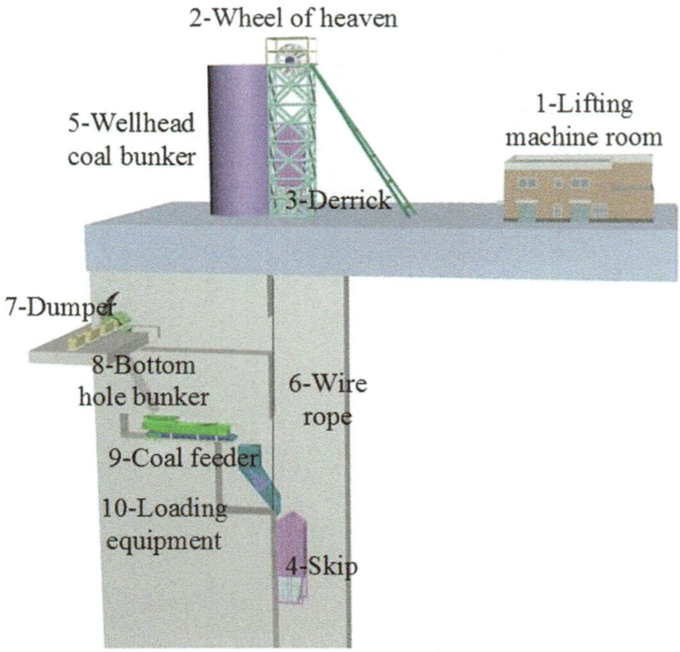

Fig. 3.17 Shaft skip lifting system

References

1. Yuan, X., Jiang, T., Su, C. et al.: The transform of 3dsmax polygon model to UG engineering model. J. Wuhan. Univ. Technol. **28**(5), 95–97 (2006)
2. Xu, Z.: Research on virtual teaching experiment system of Aeroengine. Master's thesis of Northwest University of technology, Xi'an (2007)
3. Zhou, S.: Research and implementation of fire simulation animation based on OSG. Master's thesis of East China University of science and technology, Shanghai (2012)
4. Wang, H.: The research of the Coal Industry on the ground layout and beautify of the building design. Master's thesis of Liaoning University of engineering and technology, Fuxin (2008)
5. Weijian, C.: Mine transportation and hoisting equipment. China University of mining and Technology Press, Xuzhou (2007)

Chapter 4
Virtual Assembly Method Based on OSG

4.1 Introduction

After the virtual reality resource library is completed, by using the files in the resource library, and based on Visual Studio 2010 software and OSG virtual assembly methods and systems, use C++ language and open source function library OSG (Open Scene Graph) to develop related functions, and use CEGUI (Crazy Eddie's GUI) to design man–machine operation Navigation menu. This chapter introduces the OSG-based virtual assembly method and system implementation method and process.

The OSG-based virtual assembly method and system can make the disassembly, assembly, display and training functions of various parts of the coal machine equipment. The main workflow is: build a virtual assembly environment composed of OSG and CEGUI menus, and read the virtual reality resources in the previous chapter accordingly. The corresponding files in the library are realized through programming. It is mainly divided into operation scenes and free demonstration scenes. The system framework is shown in Fig. 4.1.

- Operating scene: read each individual component in a corresponding scene and associate the corresponding coordinates. Enter the system and develop various operating functions through programming. These functions mainly include model selection, model movement, rotation and zooming, path recording, path playback, model reset, automatic positioning constraints, and network collaborative assembly.
- Automatic demonstration scene: read the already made parts disassembly and assembly animation planned according to the optimal path into the system, and then through programming, the "automatic assembly demonstration" and "automatic disassembly demonstration" can be automatically activated, and will appear the corresponded "Introduction to Parts" function.

The key technology of software system design is shown in Fig. 4.2.

X. Wang et al., *Virtual Reality Technology in Mining Machinery*,
https://doi.org/10.1007/978-981-16-4408-5_4

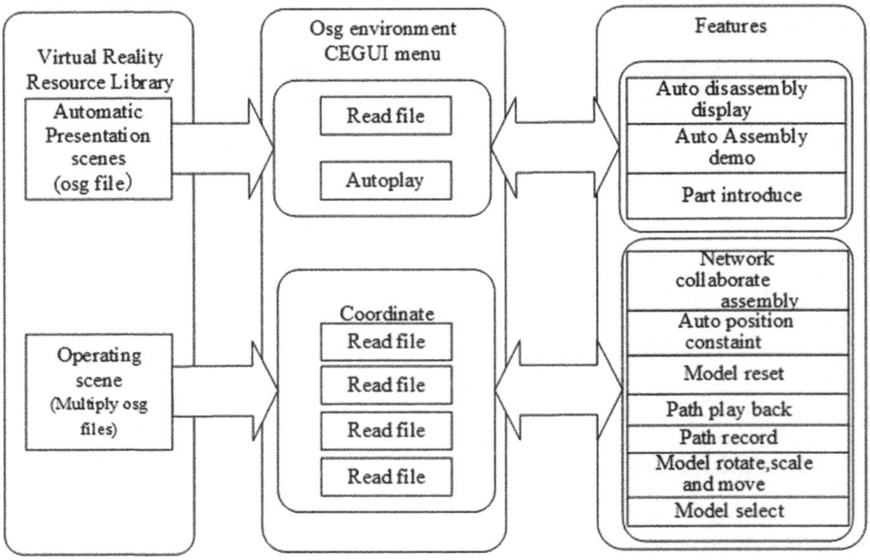

Fig. 4.1 OSG-based virtual assembly method and system framework

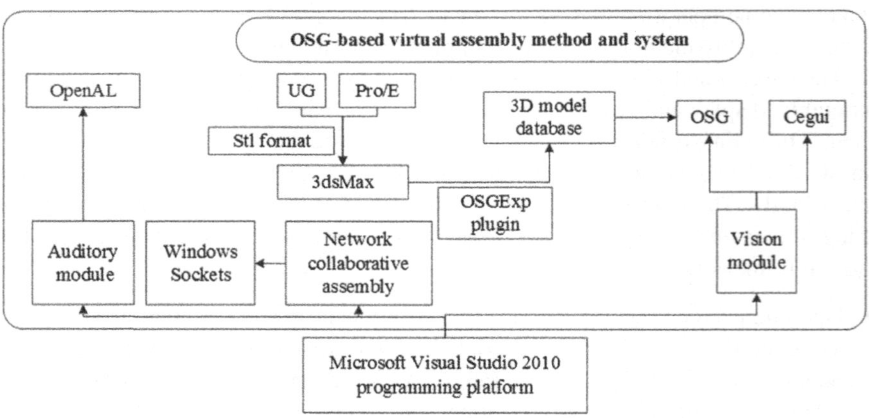

Fig. 4.2 Software system design and key technologies

4.2 Coal Machine Equipment Assembly Sequence and Path Planning Method

The assembly process planning of coal mining equipment includes assembly sequence planning and assembly path planning. Assembly sequence planning mainly studies the product assembly sequence and its geometric feasibility, finds the assembly sequence that meets the geometric and technological conditions, and

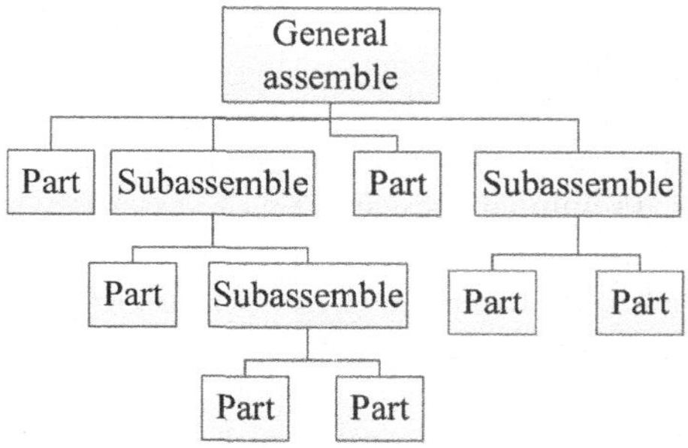

Fig. 4.3 Assembly is assembly sequence

gradually assembles the product. Assembly path planning mainly solve the path problem during product assembly, requiring each part to move to the target position according to a certain assembly trajectory to avoid interference with other parts, and the assembly path should be as short as possible [1]. In this section, the coal mining machine is still taken as an example to illustrate the coal machine equipment assembly sequence and path planning method.

This chapter uses two methods to study the assembly planning of the shearer. One is to study the assembly planning problem through the normal assembly of parts. The other is to research the assembly planning through its reverse process disassembly. The virtual assembly process and disassembly process of the shearer are reciprocal processes, and the assembly sequence planning problem is actually the disassembly sequence planning problem [2]. The virtual disassembly method assembly planning is mainly to select parts and their disassembly directions in order. The computer refines the selected parts along the disassembly direction according to a given step, and changes the spatial position, gradually disassemble each part in turn. The assembly disassembly sequence is shown in Fig. 4.3.

For the shearer, according to the assembly experience, knowledge and conventions, the experimenters perform virtual assembly (or disassembly) of the shearer model, and the system records the product assembly sequence (disassembly sequence) and assembly process (disassembly process) information, and obtain (or Reversely request) information such as the assembly sequence (disassembly sequence) and the assembly path (disassembly path) of the parts, so as to carry out the research on the assembly process planning of the shearer. Taking the parts of the shearer cutting part as an example, we will develop a virtual assembly system to conduct in-depth assembly planning research on the assembly modules of the shearer. However, the assembly sequence of the shearer needs to be researched when the system is developed. A certain preliminary study of the content, and carried out to facilitate the reasonable selection of research methods and the correct layout of the virtual scene when developing the system.

The cutting part is the part responsible for coal falling and coal loading in the shearer. Its main function is to transmit power, cut coal from the coal seam and load it into the scraper conveyor [3]. It mainly includes rocker arms, spiral rollers, gearboxes, internal spray systems, lubrication cooling systems, etc. Preliminary research on the assembly sequence of the height-adjusting cylinder of the cutting unit, and layout of it in the virtual scene, as shown in Fig. 4.4. For the piston connecting rod and piston with the label 0, the cylinder barrel with the label 1 must be installed first, and then the cylinder seat with the label 2 should be installed.

Preliminary research on the assembly sequence of the transmission gear of the cutting unit, and layout of it in the virtual scene, as shown in Fig. 4.5. First, it is necessary to assemble the first axis to the fifth axis separately, and then connect the planetary reduction device in the cutting drum with the fifth axis to complete the assembly of the entire transmission gear.

Preliminary research on the assembly sequence of the cutting part as a whole, and layout of it in the virtual scene, as shown in Fig. 4.6. There is no fixed assembly sequence for the overall assembly of the shearer cutting part. In practice, it should be further discussed and studied based on the location of the shearer.

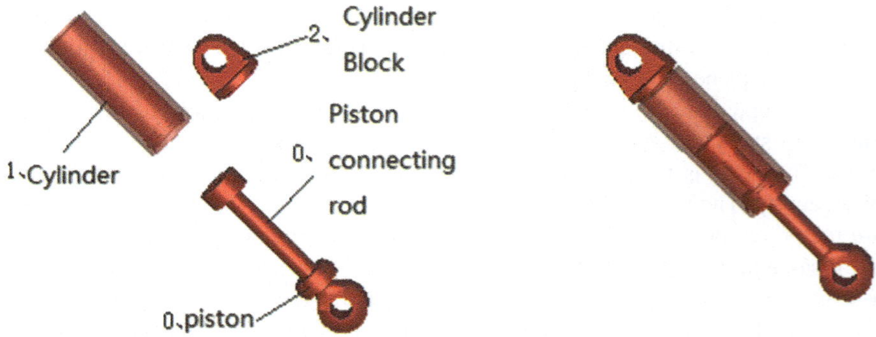

Fig. 4.4 Plan for the height adjustment cylinder assembly of the cutting department

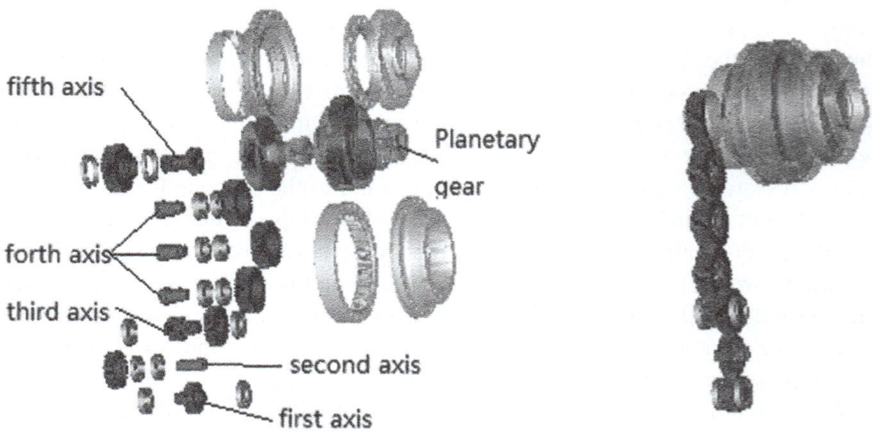

Fig. 4.5 Preliminary plan for assembly of transmission gear of cutting unit

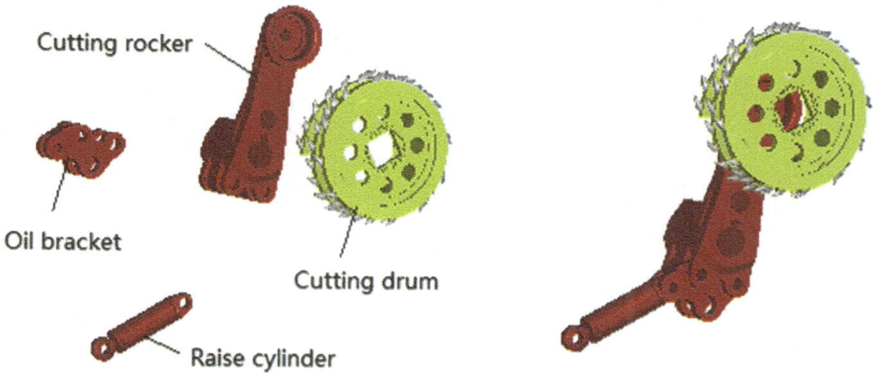

Fig. 4.6 Preliminary planning of cutting department assembly

4.3 The Framework of the Combination of OSG and CEGUI

Under the OSG virtual reality engine, it is necessary to design a user interaction interface to integrate all functions, according to their own needs, users can interact with the system. When the user interacts with the system interface, and it should be possible to complete the corresponding instructions by clicking the mouse or using shortcut keys. This section uses the combination of OSG and CEGUI to create the system operation interface.

4.3.1 Combining OSG and CEGUI to Develop System Interface

CEGUI is an open GUI library that does not charge any fees. Its existing interface mode and object-oriented programming make the entire interface development simple. In the OSG program, CEGUI is a simple drawing, which is not substantially different from the drawing of OSG's own scene. In the OSG main program, it is added to the OSG scene drawing through the addDrawable() function to realize the combined use of OSG and CEGUI, The schematic diagram of the combination of the two is shown in Fig. 4.7.

Therefore, when actually using CEGUI to develop the system interface, you should first create an interface class CEGUIDrawable which is inherited from OSG::Drawable. In this class, create the system operation interface through CEGUI and generate event response functions for the functions implemented by the system. Secondly, create an event callback class CEGUIEventCallback inherited from OSGGA::GUIEventHandler, which is mainly responsible for responding to mouse

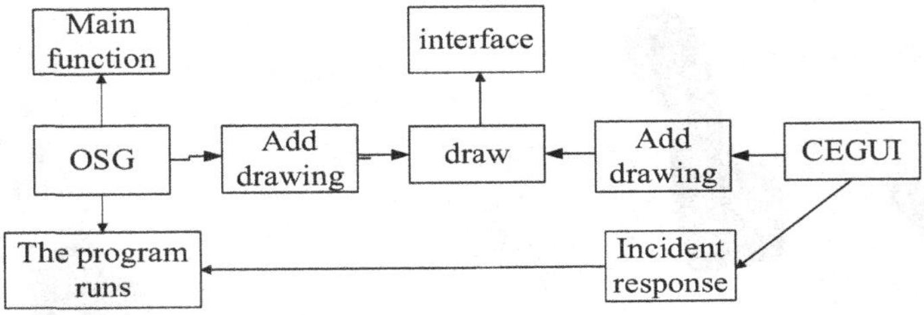

Fig. 4.7 Schematic diagram of the combination of OSG and CEGUI

operation events in the interface created by CEGUI. The two classes inherited to implement the system user interface are as follows:

```
//Implement CEGUI to create user interface
classCEGUIDrawable: public OSG::Drawable
{
};
//Responding to mouse events in CEGUI
classCEGUIEventCallback: public OSGGA::GUIEventHandler
{
};
```

It should be noted that the coordinate system of OSG on the screen is not consistent with the coordinate system of CEGUI on the screen. As shown in Fig. 4.8, the mouse position obtained in OSG is (ea.getwX(), ea.getY()), The position coordinates of the mouse in CEGUI are (ea.getX(), ea.getWindowHeight()-ea.getY()), where ea.getWindowHeight() refers to the height of the two-dimensional screen.

After completing the above two classes, the user interface function of the system can be realized by adding CEGUIDrawable, an interface class inherited from OSG::Drawable, to the scene structure tree in the main function of the system program for drawing and rendering.

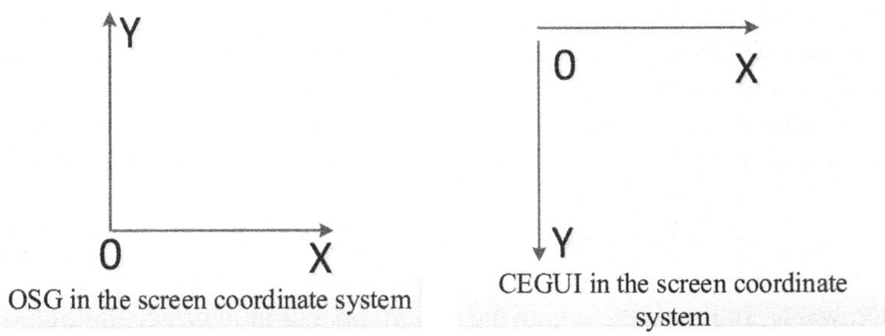

Fig. 4.8 The difference between OSG and CEGUI coordinate systems in the two-dimensional screen

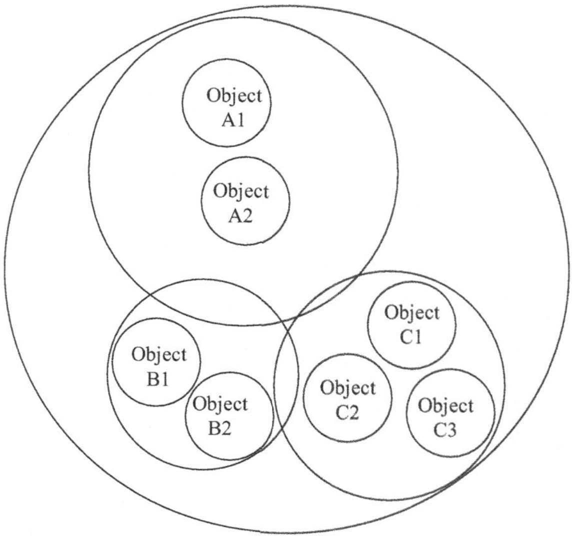

Fig. 4.9 Enclosure hierarchy diagram

4.3.2 System Scene Management

OSG uses the enclosure hierarchy to manage all virtual assembly scenes. In 3D space, the common forms of bounding volume are: bounding sphere, bounding box, bounding cylinder, K-DOP, etc. Different forms of enclosures have their own different applications. In OSG, enclosures and bounding boxes are used. In view of the actual assembly habits and the visual experience of the human eye, the system adopts the form of enclosing balls to achieve the optimal scene as much as possible. The organizational structure and the most efficient access performance, the enclosing body hierarchy diagram is shown in Fig. 4.9.

OSG uses the enclosing scene tree structure [4] level to meet the requirements of managing all virtual assembly scenes. In the scene tree structure, the Node class is used to express a basic node, including a single root node, many interrelated branch nodes and many independent leaf nodes. The root node of the tree is at the top of the entire scene tree, which is the root of the entire scene; branch nodes are used to build the level of the scene tree and implement various functions required by the system; the leaf nodes are at the bottom layer to store and manage models, Pictures, text messages, etc. The scene tree of this system is shown in Fig. 4.10.

4.3.3 Scene Menu Content Design

The menu list established by CEGUI is shown in Fig. 4.11.

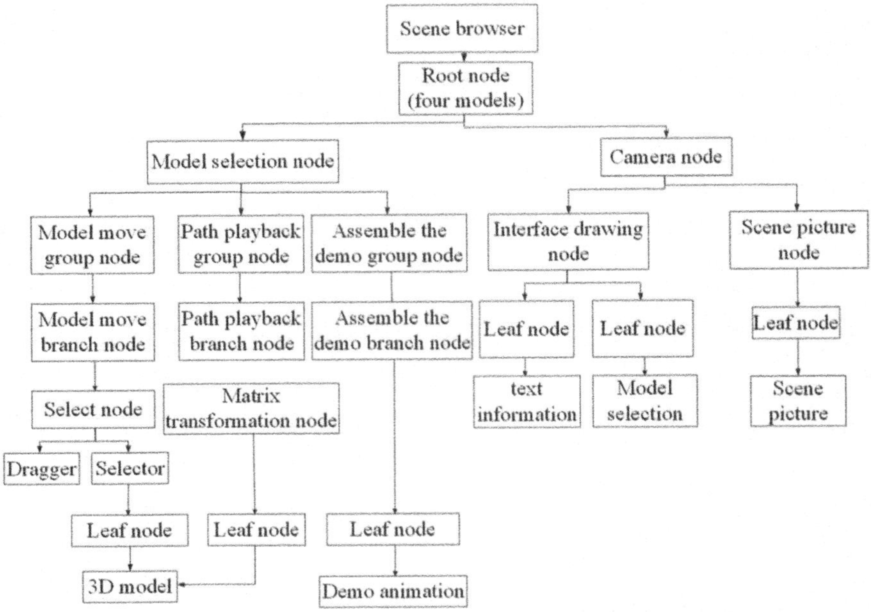

Fig. 4.10 Scene tree diagram

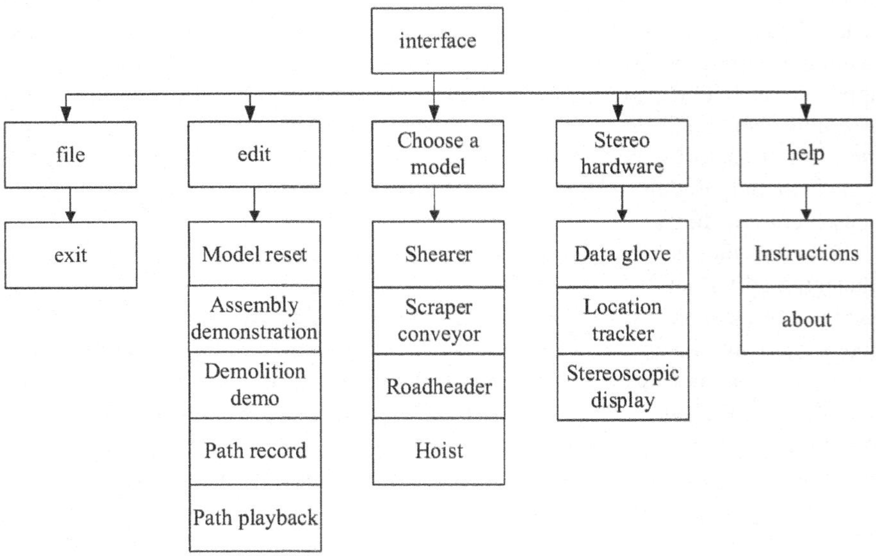

Fig. 4.11 System interface diagram

Table 4.1 CEGUI functions used in the menu bar

Types of	Functions used
Toolbar	Menu bar
Text display	Menu item
Second-level choice	Popup menu
Popup menu	Frame window
Button selection	Button
Text input	Edit box
Drop-down menu options	Multline editbox
Static text	Static text

Among them, the theme of the interface adopts the TagarezLook mode of CEGUI, and various types of menu CEGUI functions are shown in Table 4.1: The operating interface of the system allows users to conveniently select different functions required during assembly.

The operating interface of the system allows users to choose different functions according to their own assembly requirements. The specific functions are as follows:

- "File" is mainly used to exit the operation interface;
- "Edit" can switch the functions set by the system;
- "Select model" can switch the explosion scene diagram of the related combination of the four models;
- "Three-dimensional hardware" can realize the switching of various three-dimensional display effects;
- The "Help" menu explains the operating method and copyright of the system.

4.4 Model Manipulation

The manipulation of the model is the basis for the realization of other functions of the system, including the selection of the model through the mouse and keyboard, and the assembly operation of the model by moving, rotating and zooming after the selection or resetting the model to the initial state. The research of model manipulation technology belongs to the research of 3D world interaction technology, which is equivalent to a kind of "control-display", which uses 3D human–computer interaction tools to input control information to the system, and then the system outputs the execution result to the user. The manipulation of the model includes selecting the model through the mouse and keyboard, and then moving, rotating and zooming the model after the selection. After the operation is completed, the model is reset to the initial state. The operation of the model requires the use of the visitor, which can apply user-defined operations to the nodes of each model to perform node operations. It allows designers or assemblers to select the required parts for assembly operation according to their own needs, which makes reasonable use of resources and shortens the assembly time.

4.4.1 Model Selection

Roaming is the basic task of observing and manipulating the entire 3D world. Before selecting the model, first determine the roamer of the entire scene. In this chapter, we will select the default trackball roamer of OSG. In a 3D scene, the position of the object model is fixed and the rover only changes the visual observation angle, and the rover only considers the rationality of the visual movement. Trackball roaming uses the OSGGA: Trackball Manipulator class to achieve visual positioning with the help of the movement of the mouse. Click and hold the left mouse button to rotate the entire scene in a three-dimensional space like a trackball.

In addition, the way of roaming is inertial. After a certain operation is performed on the mouse, even if the mouse is released, the visual effect will continue to run in that direction. The OSG viewfinder object Viewer uses the setCameraManipulator() function to set a target for this Viewfinder robot.

The selection of the model is realized by judging the response function of the click of the left mouse button. The core that the model can be selected is that the three-dimensional coordinates of the mouse are the same as the three-dimensional coordinates of the model. The specific implementation process is shown in Fig. 4.12.

The concept of manipulating the movement, rotation and zooming of three-dimensional objects is completely different from the concept of roaming. It does not change the observer's perspective and point of view, but transmits interactive events to the user. When the object is selected, a Cartesian coordinate system will be established at its coordinate center. The movement of the object is realized by clicking the left mouse button on the XYZ axis of the coordinate system while dragging the mouse. Click the right mouse button to rotate the model, press left key and right key at the same time to zoom the scene.

In OSG, set a MatrixTransform control parent node for the object to be manipulated, and change the object that needs to be manipulated by changing the content of the transformation matrix of the parent node. The command manager method is used to implement the Dragger in the OSGManipulator library.

When the mouse sends an interactive action command to the dragger, the dragger sends the command to the command manager, and the manager distributes it to the candidate objects. The candidate objects control the nodes set by the object to achieve the control of the model.

The Matrix class is used in OSG to express a 4 * 4 matrix, which can represent the position of the object in the virtual space. On the basis of this matrix, the movement, rotation and scaling of the object can be achieved by multiplying the correlation matrix to the right.

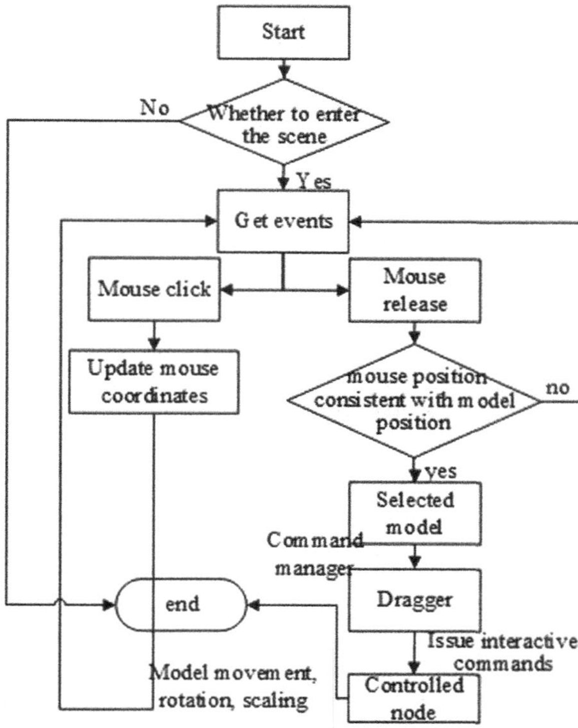

Fig. 4.12 Model selection, movement, rotation and scaling flowchart

$$T = \begin{bmatrix} 1 & 0 & 0 & 0 \\ 0 & 1 & 0 & 0 \\ 0 & 0 & 0 & 1 \\ dx & dy & dz & 1 \end{bmatrix} \tag{4.1}$$

Equation 4.1 is a translation matrix, which can translate the model along the X axis by dx, along the Y axis by dy, and along the Z axis by dz;

$$Rx = \begin{bmatrix} 1 & 0 & 0 & 0 \\ 0 & \cos\theta & \sin\theta & 0 \\ 0 & -\sin\theta & \cos\theta & 0 \\ 0 & 0 & 0 & 1 \end{bmatrix} \tag{4.2}$$

The Eq. 4.2 matrix rotates the model around the X axis by the angle θ;

$$Ry = \begin{bmatrix} \cos\theta & 0 & -\sin\theta & 0 \\ 0 & 0 & 1 & 0 \\ \sin\theta & 0 & \cos\theta & 0 \\ 0 & 0 & 0 & 1 \end{bmatrix} \quad (4.3)$$

The Eq. 4.3 matrix rotates the model around the Y axis by the angle θ;

$$Rz = \begin{bmatrix} \cos\theta & \sin\theta & 0 & 0 \\ -\sin\theta & \cos\theta & 0 & 0 \\ 0 & 0 & 1 & 0 \\ 0 & 0 & 0 & 1 \end{bmatrix} \quad (4.4)$$

The Eq. 4.4 matrix rotates the model around the Z axis by the angle θ;

$$S = \begin{bmatrix} x & 0 & 0 & 0 \\ 0 & y & 0 & 0 \\ 0 & 0 & z & 0 \\ 0 & 0 & 0 & 1 \end{bmatrix} \quad (4.5)$$

Equation 4.5 is a zoom matrix, which can zoom the model by x times along the X axis, zoom by y times along the Y axis, and zoom by z times along the Z axis.

For the model whose coordinates are matrix Y, perform a series of operations on it, and the corresponding operation can be achieved according to the following formula 4.6.

$$Y' = Y \cdot T_N \cdots T_1 \cdot Rx \cdot Ry \cdot Rz \cdot S_N \cdot S_1 \quad (4.6)$$

Y is the coordinate before moving, Y' is the coordinate after moving, and $T_N T_1 Rx Ry Rz S_N \cdots S_1$ is the transformation matrix.

In the program documentation, use the following statements to realize the movement, rotation and scaling of the model respectively.

```
OSG::Matrix mat1; // Define a matrix
mat1.preMultTranslate(); // Translate the matrix
mat1.makeRotate(); //Rotate the matrix
mat1.makeScale(); // Scale the matrix
```

4.4.2 Model Reset

In OSG, an enclosing body of the enclosing ball type is used, and the center of the enclosing ball expresses the level of the local coordinate system in the scene tree where it is located, not the center of the object in the world coordinate system. In the OSG-based virtual assembly system, each single component is contained in a bounding sphere. Therefore, before importing the model into the system, it is necessary to record the coordinates of each model in 3DMAX to use the coordinates Instead of the center coordinates of each model in the virtual scene.

In 3DMAX, there are two coordinate systems, one is the world coordinate system, and the other is the view coordinate system. When recording the coordinates, it should be in the view coordinate system, click the model to record its coordinates, and then the view set all the coordinate systems to zero and reset the coordinate axes to make the two coordinate systems coincide, and then transform the model into the virtual scene. This ensures that the center coordinates of the model in the virtual scene coincide with the coordinate system of the virtual scene, and then set the coordinates recorded in 3DMX so that the coordinate position in the virtual scene is exactly the same as that set in the 3DMAX software. On this basis, no matter what state the model is manipulated to, it can be reset to the initial state.

The process of resetting the model in the virtual scene is to go through a series of judgments as shown in Fig. 4.13. If the requirements are met, the position of the model imported into the virtual scene is reset to the initial state at the time of import.

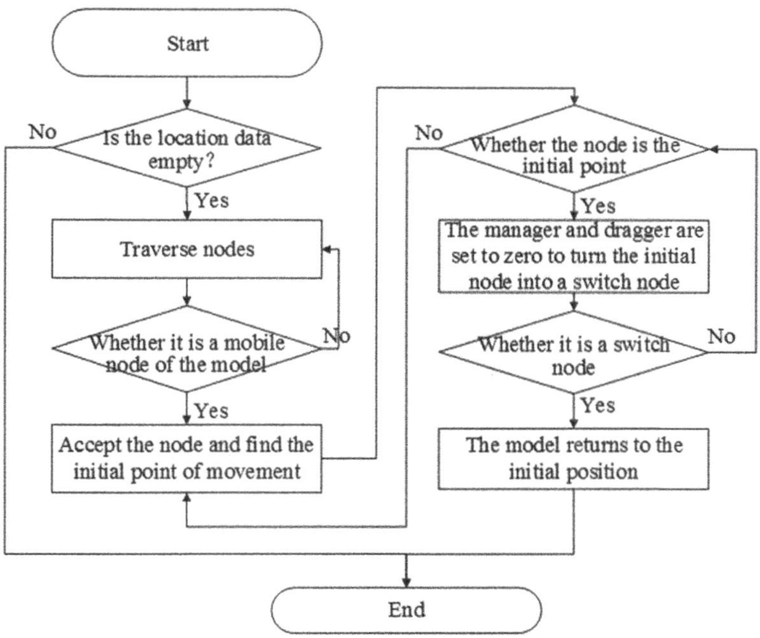

Fig. 4.13 Model reset flow chart

4.5 Virtual Assembly and Disassembly Demonstration

Automatic demonstration means that after the user selects this function, the system can automatically move the model to be assembled to the correct assembly position. Its purpose is to provide a better assembly method for experimenters to observe, compare, and analyze, and it should have a realistic demonstration effect.

The format of the virtual assembly demonstration is exactly the same as the format of the imported basic model, and its use in the entire program is similar to that of the model. The virtual assembly demonstration process is shown in Fig. 4.14.

In this chapter, frame animations are made for each module of the shearer in the 3DMAX software, and they are imported into OSG for automatic demonstration. This method is based on interpolation, and the presentation effect is smooth. When the frame animation is exported, the loop mode is set to a single pendulum (SWING),

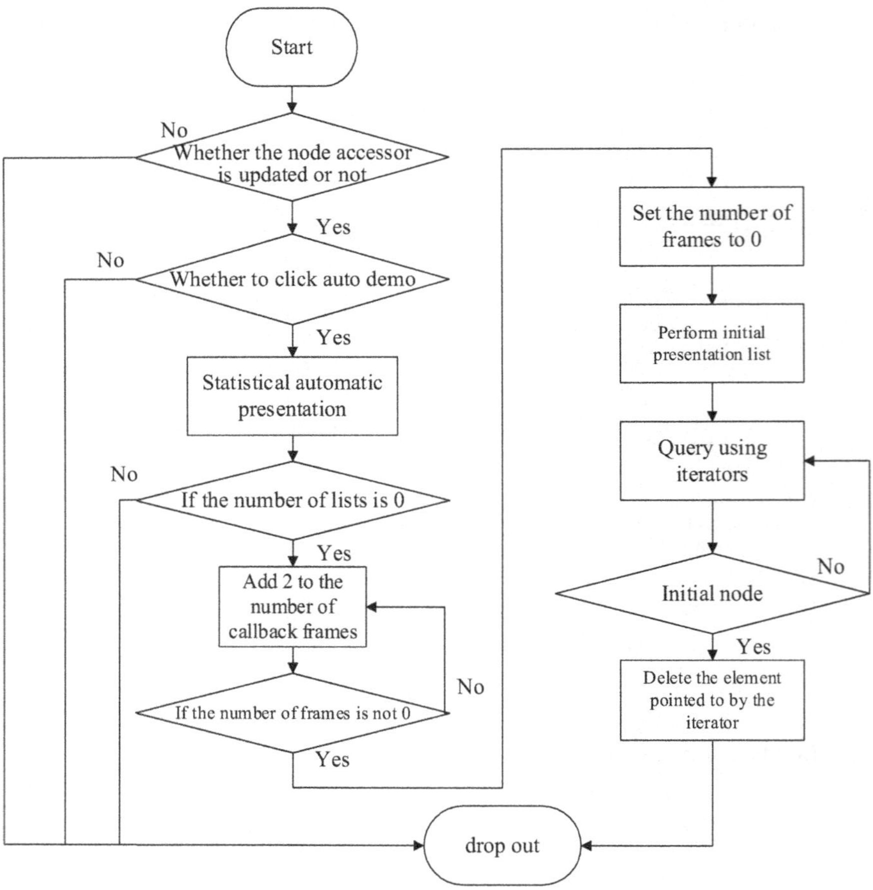

Fig. 4.14 Virtual assembly demonstration flow chart

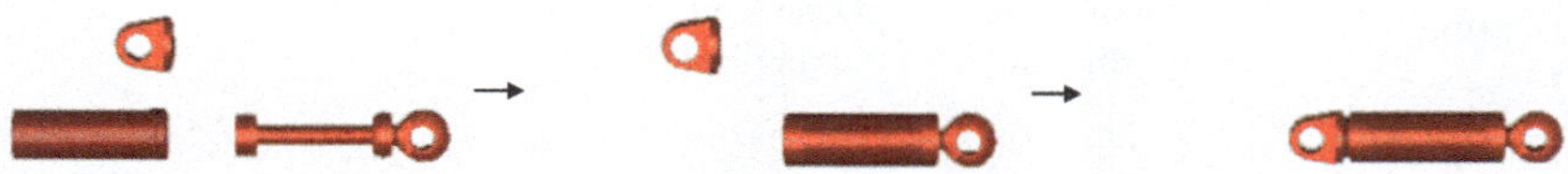

Fig. 4.15 Automatic demonstration of height adjustment of the cutting part

which can not only automatically demonstrate the assembly process of parts, but also reflect the disassembly process. When implementing this function, we mount the model with path animation made under the model automatic demonstration group node in Fig. 4.15. When the experimenter chooses the automatic demonstration, it will be displayed to provide assembly demonstration.

The virtual assembly demonstration provides users with the optimal model assembly and disassembly sequence, which reflects the assembly sequence from the initial part to the final part when the model is accurately assembled. The standardized assembly sequence lays a solid foundation for the reasonable design and precise assembly of the equipment for the operators.

Among them, this chapter obtains the best assembly sequence by disassembling and solving the assembly sequence [5]. For the transmission gear part of the cutting part, there are many parts, but the position of the transmission gear is basically arranged in a line, and the parts are easy to place in the space. The disassembly sequence and experience are used to disassemble and assemble, as shown in Fig. 4.16.

Although the virtual assembly demonstration is simple, if you want to achieve a better demonstration effect, you need to follow the steps shown in Fig. 4.17.

After importing the model into 3DMAX by the 3D design software UG or Pro/E, record the coordinates of the assembled parts according to the method in the previous section, and then disassemble the model. After disassembly, use the same method to record the scattered The coordinates of each component of the state are used to make the animation of the virtual assembly according to the different coordinates of

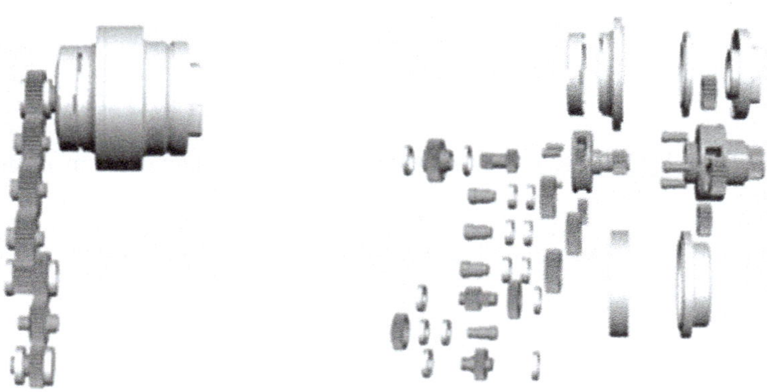

Fig. 4.16 Assembling and disassembling diagram of cutting part transmission gear

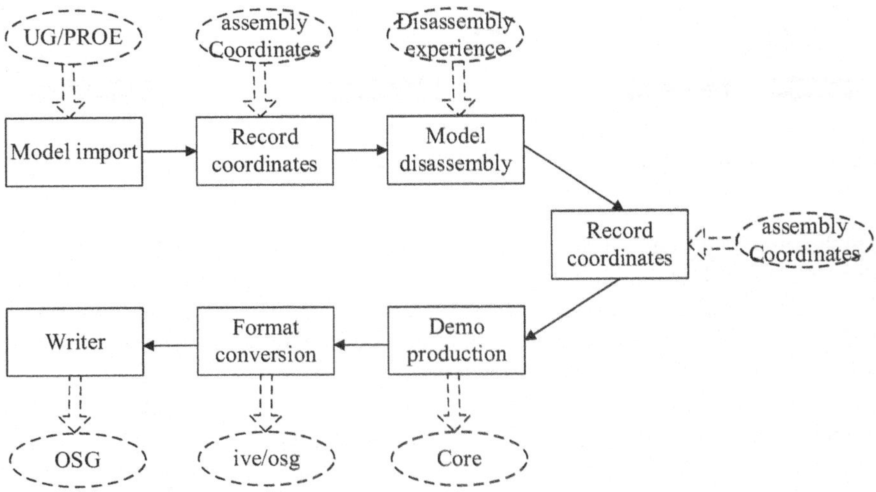

Fig. 4.17 Virtual demo flow chart

the two times, and the form of the animation is converted into OSG or IVE format and written into the main program. The detail code is as follows:

OSG::ref_ptr < OSG::Node> auto_jgb_jgtzj = OSGDB::readNodeFile ("E:\\juejinji\\jgb\\jgtzj\\jgtzj.ive");

//Enter the initial assembly animation of the system interface.

OSG::ref_ptr <OSG::Node> auto_csdh = OSGDB::readNodeFile ("E:\\tishengji\\jiansuqi\\ZZL1000\\ZZL1000-cx.ive");

//Demonstration animation of automatic assembly of parts.

createModel("E:\\tishengji\\zhuzhou\\shuangguntong\\1–1.ive", std::string("ga1"), cmd.get(), move_dansheng_zzzz.get(), record_dansheng_zzzz.get(),restorelist, OSG::Vec3 (4644.236, -2233.345, 55));

//Read a single OSG model and set the recorded coordinate position as the initial position.

Read a single OSG model and set the recorded coordinate position as the initial position.

4.6 Path Record and Playback

The assembly path refers to a series of spatial transformation processes experienced by the assembly from the initial pose to the pose when the assembly has a certain geometric positional relationship with the restraint (or the installation base) in order to avoid collision and interference [6]. The user performs multiple assembly operations on the model and uses the animation playback to discover the problems and design defects of the equipment in the actual assembly process, so as to find the best assembly path and provide the relevant personnel to the reasonable design of the equipment. And accurate assembly provides the basis.

4.6.1 Path record

In OSG, the OSG::Animation Path class is used to record the assembly operation path. When the record path function is turned on, a structure is first automatically defined to record the name of the node manipulated by the model, the time of each frame of the model's action, and the pose matrix of each time in the process of manipulating the model. Click again to close the record and transfer the recorded structure information to the vector group list. After the record list manipulated by the model is formed, the list is accessed in the form of depth-first traversal, and the txt file is generated at the specified location, and it is used for path playback function.

4.6.2 Path Playback

In the virtual scene, the path record is obtained through a series of discrete spatial location points experienced by the operator manipulating the model [7]. The playback is to re-traverse the nodes recorded during manipulation in the form of depth-first traversal. This traversal form prioritizes the longitudinal query of the nodes until the end leaf nodes, and then gradually returns to the nodes that have not been visited at the previous level. The access process is shown in Fig. 4.18.

Path playback is realized on the basis of path recording. Path playback uses the UpdataCallback function in the OSG::AnimationPathCallBack class in OSG, and is completed by adding an update callback function to the group node of the path record.

The playback of the path record of this system adopts an infinite loop mode, which can continuously playback the last recorded assembly operation until the user considers it appropriate to forcibly close it.

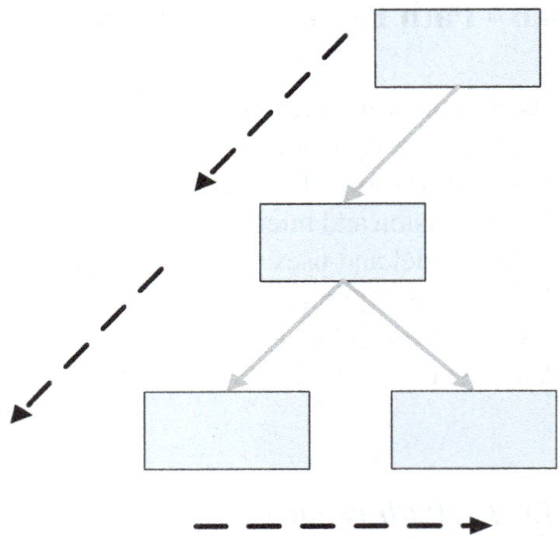

Fig. 4.18 Node traversal process diagram

4.7 Automatic Positioning Constraints

At present, precise positioning technology mainly includes geometric constraint automatic recognition precise positioning technology, interactive constraint definition precise positioning technology and target position precise positioning technology [8–10].Geometric constraint automatic recognition positioning technology means that the system automatically recognizes the geometric constraint relationship between components according to the assembly intention of the experimenter.

In the OSG-based virtual assembly method and system, the user manipulates the model through the mouse and keyboard; however, in the three-dimensional space, the sense of pose is poor, and the assembly of the model based solely on the user's visual observation and feeling has great ambiguity and uncertainty. It is likely that there will be a gap during assembly, and it is difficult to completely assemble the model to the correct position [11].Therefore, a certain precise positioning method must be used to determine whether the components are installed in place. Currently, there are two commonly used precision positioning methods as follows:

The assembly positioning algorithm adopts the positioning algorithm collision detection and the approximate capture of the pose. According to the actual requirements of the system and the characteristics of OSG, combining the concept of assembly positioning algorithm based on collision detection and pose approximation capture, the method of automatically positioning and guiding the restraint of the surrounding ball, the principle is shown in Fig. 4.19.

The entire sphere surrounds the assembly object. When the distance between the centers of the two surrounding spheres is less than the preset distance, the model is automatically placed according to the set target information to obtain an accurate pose. The mathematical expression is:

$$\sqrt{(x1 - x2)^2 + (y1 - y2)^2 + (z1 - z2)^2} \le d \tag{4.7}$$

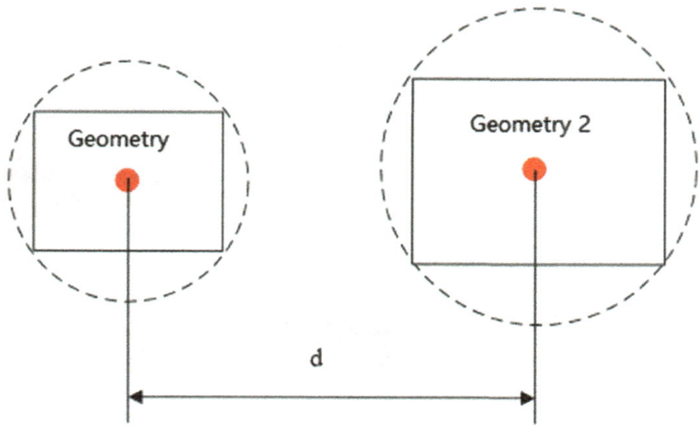

Fig. 4.19 Schematic diagram of bounding ball positioning constraint

In the formula, geometry 1 space coordinates: (*x1*, *y1*, *z1*); geometry 2 space coordinates: (*x2*, *y2*, *z2*); set distance value: d. Due to the large size of the equipment model, here we set d to 200, which can basically satisfy the user's normal visual operation in this system.

In addition, the three-dimensional space of the OSG-based virtual assembly method and system has a broad and unlimited space, and the coordinate data equipped in the three-dimensional space is possible from zero to tens of thousands, it is impossible for the user to correctly assemble the model in place after only one operation during assembly. Therefore, when performing continuous operations on the model, Need to be guided to continue the operation. This system uses three-dimensional coordinate guidance, when the center distance of the enclosing ball is less than the set distance, every time the model is operated to a position, the three-dimensional coordinate difference from the correct assembly position will be displayed. The operator can perform further operations on the model according to the prompt. It saves assembly time and ensures the accuracy of coordination. The coordinate difference calculation method is as follows:

$$\Delta d = d_1 - d_2 \tag{4.8}$$

d is the distance difference coordinate, d1 is the current position coordinate, and d_2 is the target position coordinate.

The first two of the three methods require more complex algorithms, and the third method is intuitive, clear, and not complicated to implement. Therefore, this chapter realizes the precise positioning of system assembly components on the basis of this method. First, the location of the shearer model needs to be planned in 3DMAX. Specifically, the center of each model of the shearer is placed at the origin of the world coordinate system and imported into the OSG. And record its position in the scene when it is correctly assembled and when it is not in the scene,

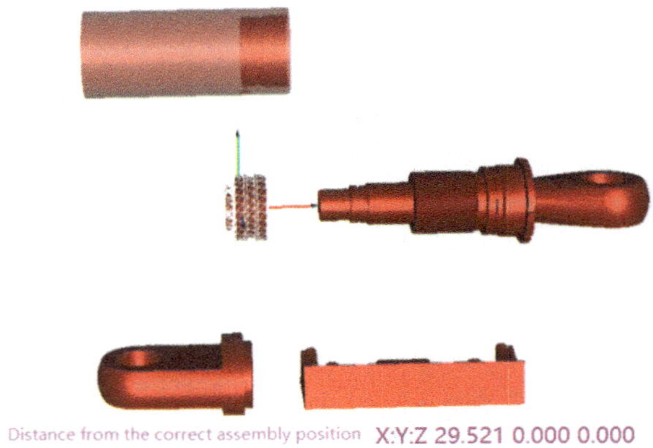

Fig. 4.20 Prompt diagram of correct assembly position

Then by using the candidate object OSGManipulator::Selection to control the position state of the assembly model under it, when the model is imported, the position information of the unassembled model is synchronously imported into the candidate object, realize the preliminary layout of the scene model. After the model is moved, use the distance formula of two points in three-dimensional space to find the distance between the current position of the assembly model and the correct assembly position. If it is less than the set value of the system, Import the correct assembly position information into the candidate object. Realize direct and precise positioning based on target distance constraints. The main functions implemented are as follows:

(1) Ach time the assembly model is moved, the distance between the current position of the assembly model and the correct assembly position will be prompted, as shown in Fig. 4.20.
(2) After moving the model every time, the distance between it and the correct assembly position should be judged, if the distance is less than a certain value, it should be located directly through precise positioning technology, and prompted that the assembly is correct. As shown in Fig. 4.21.

4.8 Network Collaborative Assembly

4.8.1 Work Process

Network collaborative design is mainly used to realize real-time synchronization operations in different places. Its core is the real-time sharing of information. In this system, it refers to the real-time transmission, reception and update of the location of the assembly model.

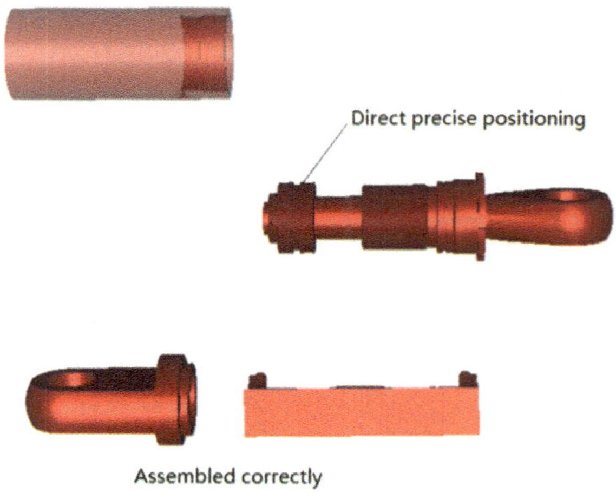

Fig. 4.21 Precise positioning of components based on target position constraints

When working, the network communication module of this system can record the information of the operated parts, including the name, the position matrix of the parts, etc., and upload it to the server, and the server transmits it to other users.

After receiving the message, the client parses and extracts the corresponding data information, and then the scene updates the position information of the operated part under the action of the callback function. The specific workflow is shown in Fig. 4.22.

4.8.2 Network Cooperative Assembly Based on Windows Sockets

Windows Sockets is a network programming specification under Windows. When developing this module, the send() and Revc() functions in Winsock are used to send and receive location information.

After the system receives the information, it needs to record and extract the information. In this section, an update queue is used for scene update and data recording. For this multi-threaded data sharing process, which one of the many operators has the right to operate? It is important, otherwise there may be a situation where multiple operators operate at the same time and cause the network and the system to collapse, so the shared resources must be shackled. When the scene is updated, a Mutex is locked. When the communication thread accesses this critical resource, it needs to wait until the update callback releases the critical resource. This avoids the problem of program errors caused by simultaneous operations of multiple users.

The following code implements the locking of the program:

EnterCriticalSectio(&Critical_Section);
Write or read data to the update queue:

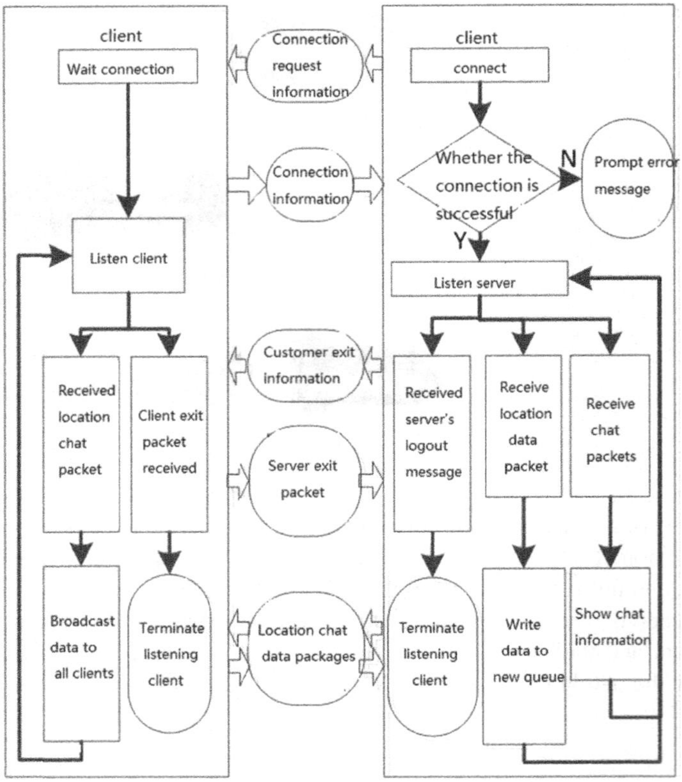

Fig. 4.22 Network collaboration workflow

LeaveCriticalSectio(&Critical_Section);

When the system reads the required location information, it must look for the model corresponding to the information, traverse and visit each node through the node visitor (OSG::NodeVisitor), find the required node, and then position the model Transform and update, thus realizing network coordination.

4.8.3 Specific Implementation Process

Collaborative assembly of local area network can achieve the same purpose of users in different regions. It will expand and increase the application scope of the virtual reality assembly system for coal mining machinery equipment. Its components include the main control terminal, server and client. The main control terminal contains all the functions and required materials of the system, that is, the virtual assembly system platform built in this system; the server functions as a transfer

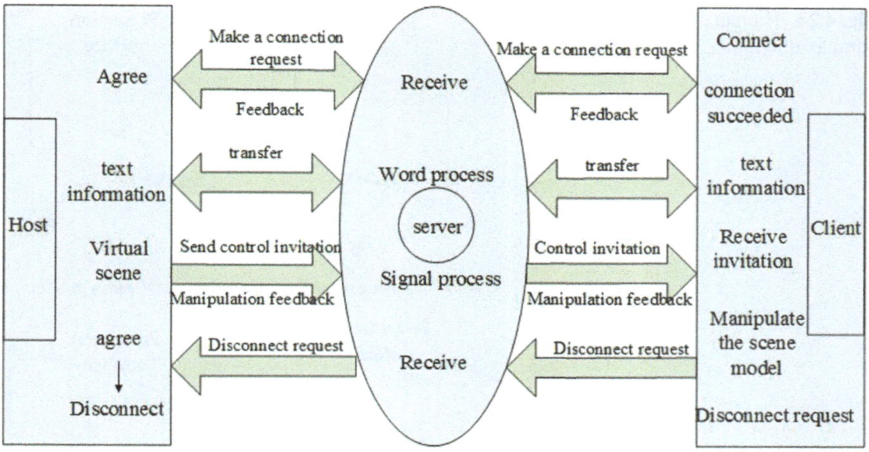

Fig. 4.23 Information flow diagram of collaborative assembly

station, transmitting and coordinating the exchange information between the main control terminal and the client; The client does not need to install the entire complex system, as long as there is a corresponding program installation package that can transmit information with the server. The information flow chart of cooperative operation is shown in Fig. 4.23.

4.9 Stereoscopic Display

Stereoscopic display is the core technology of the visual display of the virtual reality assembly system for coal mining machinery equipment. Using this technology, the position, level and depth of the system image can be displayed more vividly, and the user can more intuitively understand the real situation of each part of the mining and transportation equipment model [12].

4.9.1 Details and Types

When observing a certain thing, from a physical point of view, given that the left and right eyes are in different positions, the observation of the same thing will have a slight parallax in the retina, which causes the actual position and angle of the figure we see to be different. A certain level of parallax creates a sense of depth of observation. According to the positions of the projection surface, the binoculars and the observation object, the types of parallax that can be formed by the binoculars are shown in Fig. 4.24:

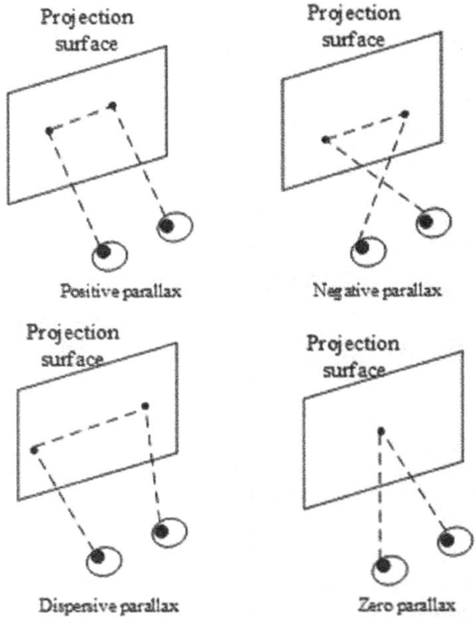

Fig. 4.24 Human eye parallax diagram

The binocular images with horizontal parallax are fused by the brain to form stereo vision in the human brain [13, 14], where the left image projected by the projector is passed to the left eye, and the right image projected by the projector is passed to the right eye, And then use the fusion machine to fuse the images formed by the left and right eyes to form a stereoscopic image with a sense of depth. As far as current technology is concerned, there are several types of 3D display that can be realized as shown in Fig. 4.25.

In this system, based on the development of stereoscopic display technology and existing hardware conditions, this system applies passive stereoscopic display technology that wears polarized glasses.

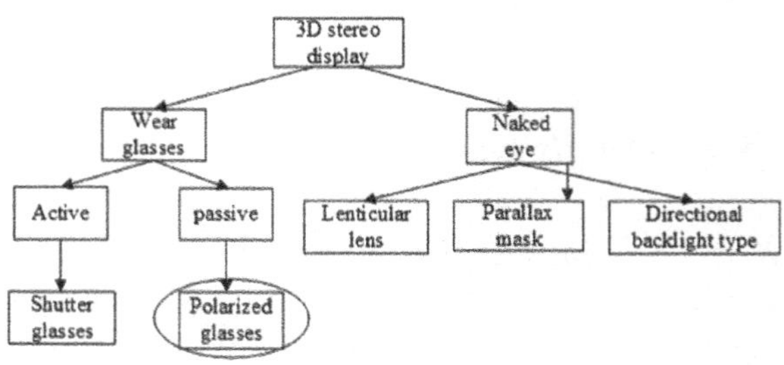

Fig. 4.25 3D stereo display type diagram

4.9.2 Method for Realizing Binocular Parallax Stereo Display

There are many types of stereo vision display methods, such as stereoscope method, chromatic aberration method, polarized light method, grating method, alternating frame method, laser holography method and so on. The virtual reality equipment and simulation system oriented to coal mining machinery equipment are realized by the polarized light method. The realization principle is shown in Fig. 4.26.

The left and right eye image signals are sent by the graphics workstation to the projector. A polarizer is added in front of the projector lens to divide the signal into horizontal and vertical signals. The signal is projected on the screen and refracted to the human eye. The user wears polarized stereo glasses. There is an analyzer in front of the glasses, the left-eye image can only pass through the left-eye lens, and the right-eye image can only pass through the right-eye lens, so that parallax appears in both eyes, thus producing a three-dimensional sensory effect. In addition, the signal will intersect in the middle of the screen and will be distorted in the arc-shaped part. Here, a fusion machine is used to perform edge fusion and nonlinear distortion correction on the four-channel projection system, so that the entire screen can display better results.

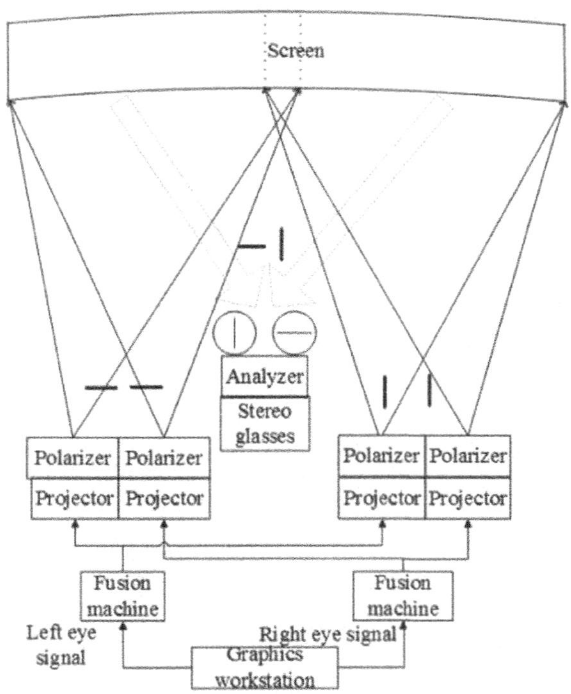

Fig. 4.26 Three-dimensional display principle diagram

This chapter is based on the binocular parallax technology to realize the passive stereo display of the experimental system. For interactive observation, it is necessary to wear hardware equipment such as polarized glasses. Binocular parallax stereo display technology essentially has the following three aspects:

- Generate a pair of images with left and right parallax;
- Make the user see a piece of image information with both eyes through the principle of polarization;
- Through the fusion of the brain's visual nerves, a three-dimensional visual effect is produced.

Specifically, the image is projected onto a large screen by using a projector and a polarizer to generate light beams with mutually perpendicular polarization directions. The user uses polarized glasses (analyzers) with mutually orthogonal light transmission directions to observe, so that the binocular parallax appears and immersion occurs.

4.9.3 Stereoscopic Imaging Technology in OSG

After studying the principle of the three-dimensional display, it is necessary to implement the three-dimensional display effect in the virtual reality assembly system for coal mining machinery equipment. In this regard, OSG encapsulates the relevant settings of the stereo display in the OSG::DisplaySettings class, and sets the switch of the stereo display through the Set Stereo (bool on) function, and can adjust the effect of the stereo display. The class is shown in Fig. 4.27.

The main program code for stereo display setting is as follows:

//Create an instance of the display setting class

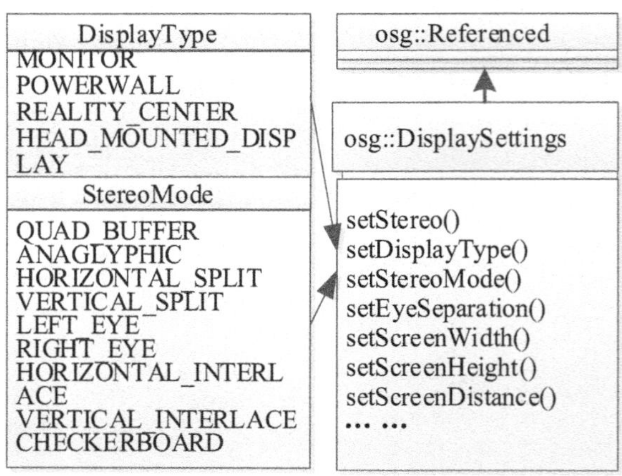

Fig. 4.27 Three-dimensional display class diagram

```
OSG::ref_ptr < OSG::DisplaySettings > displaysetting = new OSG::DisplaySettings;
//Associate the display setting class with the scene (viewer is the scene browser OSGViewer::Viewer)
viewer- > setDisplaySettings(displaysetting);
//Turn on the stereo display
displaysetting- > setStereo(true);
//Set the stereo display mode (four buffers), type (large screen projection)
displaysetting- > setStereoMode(OSG::DisplaySettings::QUAD_BUFFER);
displaysetting- > setDisplayType(OSG::DisplaySettings::POWERWALL);
//Set the distance between the eyes
displaysetting- > setEyeSeparation(eyespace);
//Set the screen width, height, distance
displaysetting- > setScreenWidth(screenwidth);
displaysetting- > setScreenHeight(screenheight);
displaysetting- > setScreenDistance(screendistance);
```

These settings should correspond to the three-dimensional display parameter setting interface designed in the system, so that the experimenter can adjust it, as shown in Fig. 4.28.

There are nine stereoscopic display modes in OSG, as shown in Table 4.2.

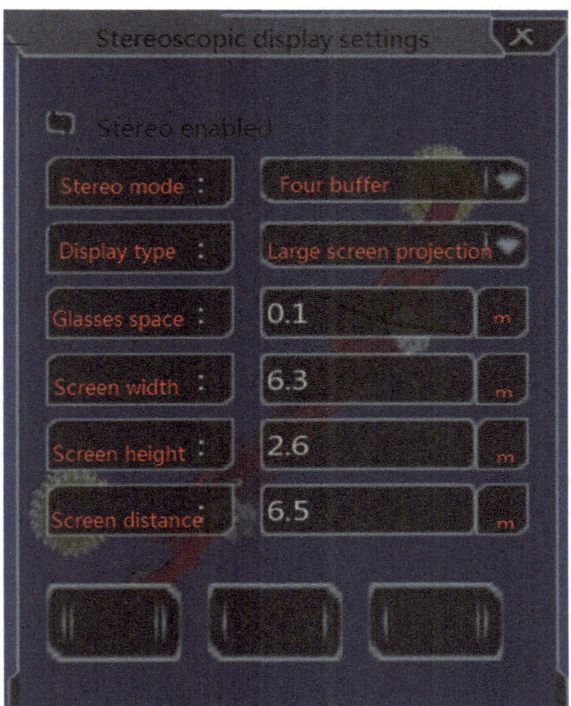

Fig. 4.28 System stereo display setting interface

Table 4.2 Stereoscopic display mode diagram

Quad buffer	Anaglyphic	Horizontal split
Vertical split	Left eye	Right
Horizontal interlace	Vertical interlace	Checkerboard

Table 4.3 Stereoscopic display type diagram (Stereoscopic display type diagram)

Monitor	Powerwall
Reality center	Head mounted display

Table 4.4 Stereoscopic display settings

Stereo mode	Four buffer (m)	Display type	Screen projection (m)
Eye spacing	0.1	Screen distance	6.5
Screen width	6.3	Screen height	2.6

There are four display types, as shown in Table 4.3:

This system uses a dual-channel cylindrical stereoscopic display projection system. Through many experiments, the various display settings are shown in Table 4.4.

In the related settings of stereo display. Glasses spacing is the main factor that affects the stereoscopic display effect. But in reality, because everyone's glasses have different spacing. In order to achieve the ideal visual effect. It is necessary to fine-tune the spacing of the glasses.

In order to facilitate the user to adjust the ideal three-dimensional effect, avoid adjusting the distance between glasses through the user interface again and again, and supports dynamic adjustment. This chapter inherits an event handler (OSGGA::GUIEventHandler), support the keyboard plus (KEY_KP_Add) and minus key (KEY_KP_Subtract) to adjust the distance between glasses. After testing, it is ideal that the size adjusted each time is about 0.05.After adjusting the distance between the glasses. You can observe a more ideal stereoscopic display effect. As shown in Fig. 4.29.It can be seen that the model display in the figure is fuzzy. According to the analysis and research on binocular parallax. This is caused by the superposition of two images with left and right parallax. When the experimenter puts on the polarized glasses to filter and fuse, they can enjoy the immersive stereoscopic display effect.

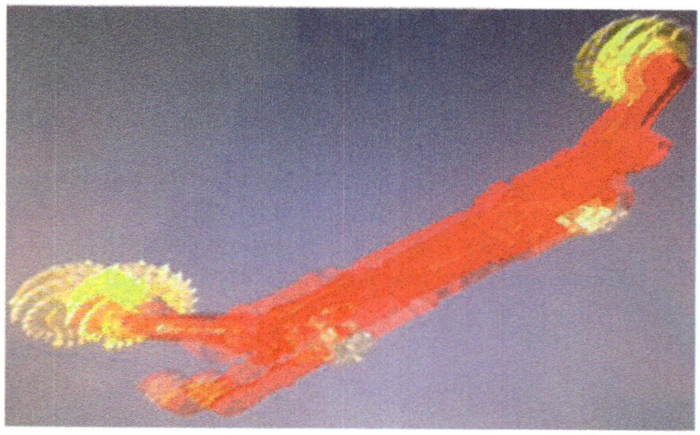

Fig. 4.29 The three-dimensional display effect of the shearer

4.10 Summary

This chapter proposes a virtual assembly method based on OSG, manipulate the model, Path recording and playback, model reset, automatic positioning constraints, Network collaborative assembly, the specific assembly methods such as the introduction of parts and automatic disassembly and assembly demonstration have been theoretically studied and the corresponding functions have been realized through programming. Based on this method, a virtual assembly system based on OSG is established, which can meet the requirements of users for virtual assembly.

References

1. Wang, F., Liao, W., Guo, Y., Wang, X., Gao, Y.: Research status and its perspective of key techniques for cable harness virtual assembly. China Mech. Eng. **27**(06), 839–851 (2016)
2. Qiang, X., Wenlei, S., Jin, Z.: The research and realization on virtual assembly of wind-generator set. Renew. Energy Resour. **28**(4), 17–20 (2010)
3. Gou, Y.: The virtual prototype and CAE analysis of the rocker arm of the shearer cutting part. Master's Thesis of Taiyuan University of Technology, Taiyuan (2010)
4. Hao, C., Teng, B.: Static Analysis for a non-uniform flexible mooring cable system. J. China Offshore Platform. **18**(4), 18–21 (2003)
5. Zhai, G.: Multi-body dynamics research on the key components of the shearer's crushing device. Master's Thesis of Taiyuan University of Technology, Taiyuan (2011)
6. Jihong, L., Junfeng, W.: Complex product collaborative assembly design and planning. Huazhong University of Science and Technology Press, Wuhan (2011)
7. Liu, J., Ning, R., Wan, B., et al.: Research of complex product assembly path planning in virtual assembly. J. Syst. Simul. **19**(9), 2003–2007 (2007)
8. Jianhua, L., Jun, Y., Ruxin, N.: Research and realization of collision detection algorithm in virtual assembly environment. J. Syst. Simul. **16**(8), 1775–1778 (2004)
9. Jian-hua, L.I.U., Ru-xin, N.I.N.G., Jun, Y.A.O., Cheng-tong, T.A.N.G.: Research on exact placement technology of component in virtual assembly environment. Comput. Integr. Manuf. Syst. **11**(4), 498–502 (2005)
10. Liu, J.H., Yao, J., Ning, R.X.: Research on information integration technology between CAD system and virtual assembly system. Comput. Integra. Manuf. Syst. **11**(1), 44–47 (2005)

11. Hu, F., Chen, C., Liu, Y.: A product assembly virtual planning system. Mach. Des. Manuf. **12**, 180–183 (2016)
12. Smith, R., Day, A., Rockall, T.: Advanced stereoscopic projection technology significantly improves novice performance of minimally invasive surgical skills. Surg Endosc. **26**(6), 1522–1527 (2012)
13. Zhao, Y., Lei, Z., Lei, M.: Research on the technology of stereo display in virtual reality. Electron. Des. Eng. **17**(10), 116–118 (2009)
14. Li, Y., Su, P., Ma, J., Mao, L.: Evaluation method and system of stereoscopic projection quality. Chin. J. Liq. Cryst. Displays **27**(1), 31–37 (2012)

Chapter 5
Virtual Assembly Method Based on UG

5.1 Introduction

Assembly is a link that takes the longest time and consumes the most labor in the coal machine equipment cycle with complex structure and poor working conditions. It is very necessary to plan the virtual assembly of the coal machine equipment virtual prototypes before the coal machine equipment is put into production. It can also solve many assembly problems that may occur in the later stage as soon as possible, shorten the cycle of coal mining equipment assembly design, reduce costs, reduce rework rates, and improve coal mining equipment assembly production efficiency. Based on the above factors, this chapter introduces the virtual assembly method and system of coal machinery equipment based on UG. This chapter takes the shearer in the coal mining equipment as the research object and develops a virtual assembly system for the coal mining machine based on UG.

5.2 Overall Design of Virtual Assembly System of Coal Shearer Based on UG

5.2.1 System Design Goals

The main design goals of the system are as follows:

(1) Reconstruct and transform the assembly model
Reconstruct the virtual assembly-oriented hierarchical structure of the shearer UG assembly model imported into the system to form an assembly tree structure for the virtual assembly process, and convert the shearer UG assembly model into a virtual assembly suitable for assembly planning and process dynamic simulation model.

X. Wang et al., *Virtual Reality Technology in Mining Machinery*,
https://doi.org/10.1007/978-981-16-4408-5_5

(2) Realize man–machine interactive assembly planning
The position and posture, assembly/disassembly sequence and path of each component can be operated by man–machine interactively, which can combine the design experience of the equipment R&D engineer and the efficient computing ability of the software system.
(3) Dynamic demonstration of the entire assembly process
The entire process of assembling and disassembling the equipment model is displayed in a visual form, and virtual simulations such as the pose display of each part and the animation simulation of the assembling and disassembling process can be realized separately. It enables R&D personnel to make overall planning and timely adjustments to the entire assembly process in advance.
(4) Plan the assembly sequence and assembly path
The virtual assembly system can plan the assembly and disassembly sequence of the shearer product and the path of each action, and can output it in text format, providing basic data support for the production of product production process cards.
(5) Good scalability
The design of the system must provide an open development environment for later system upgrades and the addition and reduction of system modules, so that the upgrades of system do not need to redesign the system, and other structures and modules of the system will not be affected when system functions are increased or decreased.

5.2.2 System Overall Structure Design

(1) System Architecture

The virtual assembly system of the shearer based on UG is mainly designed for the evaluation and analysis of the assembly performance of the shearer in the design and development stage. As the product assembly models are all CAD models established by CAD software, the system must have a data model reconstruction conversion module. In addition, according to the needs of the virtual assembly system, this chapter also designs functional modules such as assembly information display, assembly planning, assembly evaluation and assembly sequence, assembly path planning, etc.

According to the above requirements, the system framework of the UG-based coal shearer virtual assembly system designed in this chapter is shown in Fig. 5.1. The entire virtual assembly system is developed in UG software. Taking into account the upgrade of the system and the adjustment of system function modules according to actual needs, the system is designed as an open mode, and the ellipsis indicates the scalability of the system.

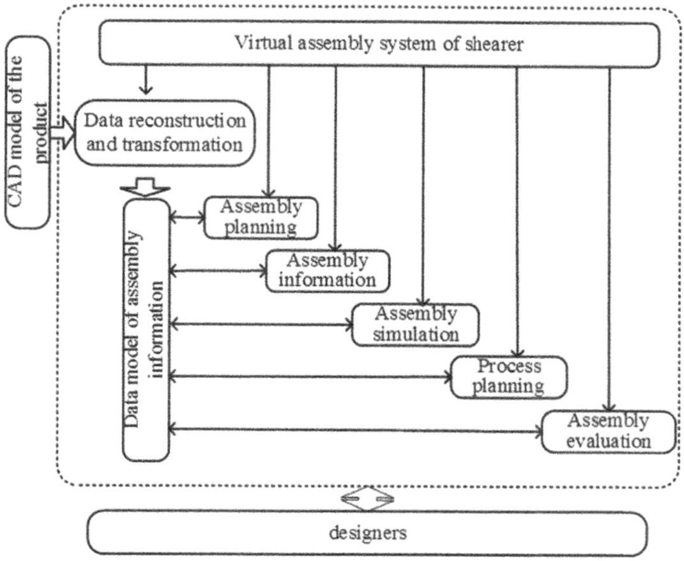

Fig. 5.1 Frame structure diagram of the virtual assembly system of coal shearer based on UG

On the human–computer interaction interface, the interaction between the designer and the system is mainly through external devices such as mouse, keyboard, and monitor.

(2) System structure design

The structural design of the system is shown in Fig. 5.2. It is divided into three modules, namely assembly model import module, assembly planning module and automatic assembly module. The assembly import module is mainly used to import the assembly model of each sub-component of the shearer and the whole machine model into the virtual assembly system environment. The structure of the assembly planning module is mainly divided into the assembly sequence planning part and dynamic simulation part of assembly process. The former includes assembly sequence and assembly path planning functions. The latter is the visual display of the assembly and disassembly process. The last automatic assembly module is mainly composed of assembly information data XML document initialization and automatic assembly.

(3) System software and hardware selection

Choosing existing CAD software platform as the development environment of the system. Choosing UG software is due to its mature application in the secondary development function and the powerful function of its own assembly environment. The hardware selection consists of a workstation computer and external devices consisting of a mouse, keyboard, and monitor. The software selection is composed

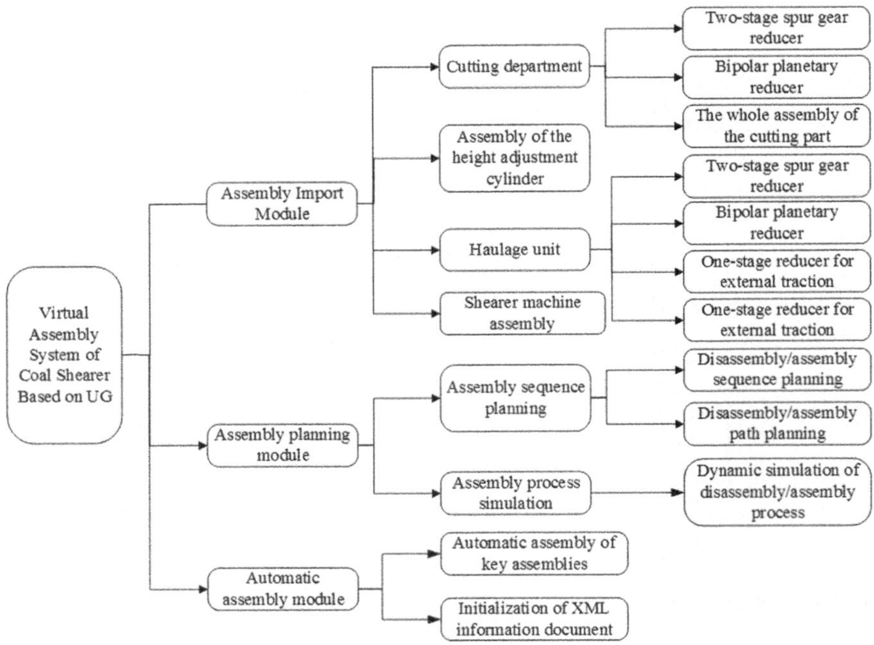

Fig. 5.2 The virtual assembly system structure diagram of the shearer based on UG

of 64-bit Windows 7 Professional Edition system and large-scale computer-aided design software UG 7.5 version.

5.2.3 System Development Environment Selection

(1) System development technology

The development environment based on the existing large-scale CAD software platform has obvious cost advantages. Existing commercial computer-aided design software, such as UG, Pro/E, SolidWorks, etc., have powerful structural design functions and provide corresponding secondary development tools. And this kind of computer-aided software is widely used and has strong versatility. Taking this as development environment has low hardware requirements, low cost investment, wide applicability and good scalability.

(2) Development language selection

UG/Open secondary development tool cooperating with VC++ development language was chosen as the development tool. VS2010 was chosen as the C++

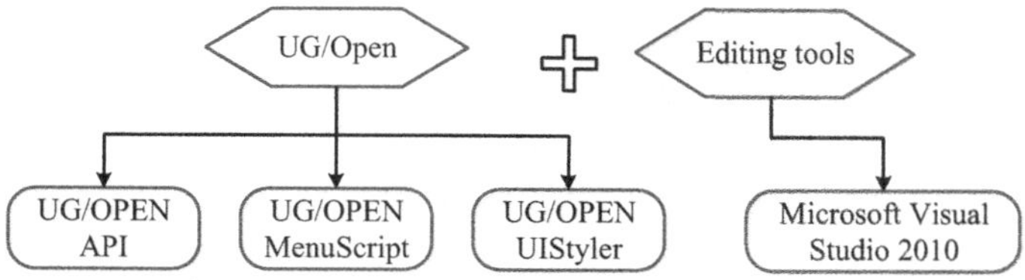

Fig. 5.3 Selection of system development language tools

language programming tool. Compared with VC++ 6.0, the feature most suitable for this chapter is that it can develop 64-bit applications.

(3) System software support

As a CAD/CAE/CAM/PDM integrated software system, UG not only has powerful solid modeling and other common functions, but also provides UG/Open secondary development tools. Using UG/Open, designers can develop a series of dedicated software systems based on UG software to meet corresponding requirements.

UG/Open is the general name of all UG secondary development tools. It is a development tool specially provided by UG software for designers and developers [1]. UG/Open development tools mainly include several modules such as GRIP, API, Menu Script, UI Styler, GRIP NC and C++. The system development tools selected in this chapter are shown in Fig. 5.3.

5.2.4 System Function Design

(1) System analysis process

In this chapter, the analysis process of the system is shown in Fig. 5.4 according to the discussed design goals.

In Fig. 5.4, the solid arrows represent man-made operation commands, and the dashed arrows represent the data interaction between the various modules in the system. According to the design structure of the shearer system, it is divided into three modules. The model processing module is mainly to import the planned assembly object model into the system according to the needs of the designers and to converse and reconstruct models to meet the needs of system planning. The dynamic planning module for assembly dynamic planning is mainly based on the virtual assembly model, providing functions such as assembly sequence planning, assembly path planning, and dynamic simulation. The static planning module is mainly composed of the display and update of assembly information data and automatic assembly functions.

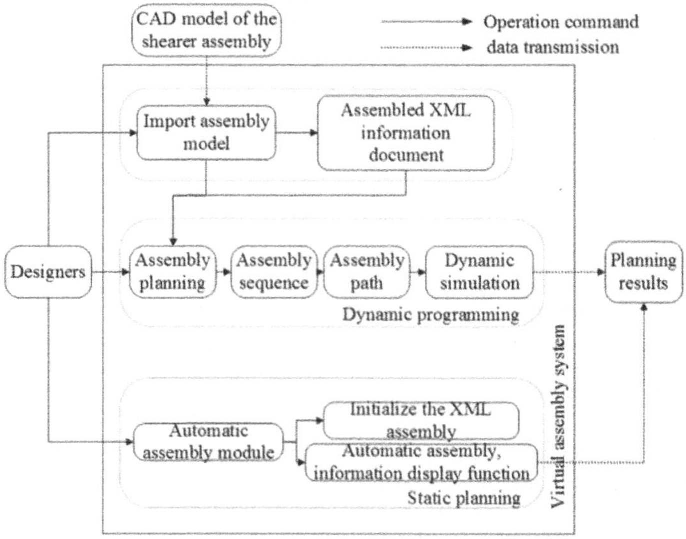

Fig. 5.4 The analysis flow chart of the virtual assembly system of the shearer based on UG

(2) Main functions of the system

This system needs to be able to read and integrate the assembly information data in the product equipment assembly model, such as the assembly relationship of all parts in the assembly, the matched geometric elements, the assembled objects and other assembly information data. It can form the data model of the entire virtual assembly system.

The main functions of the UG-based coal shearer virtual assembly system are as follows:

- Reconstructing and conversing the shearer CAD model to build a data model of the virtual assembly.
- Automatic assembly function.
- Planning of assembly sequence and path.
- Dynamic simulation of the assembly process.

5.3 Data Model Construction Method of Virtual Assembly System

5.3.1 Virtual Assembly Model Construction Method

Through legal assembly operations, the parts of production model can form information model which can express their position and attitude information. People combine

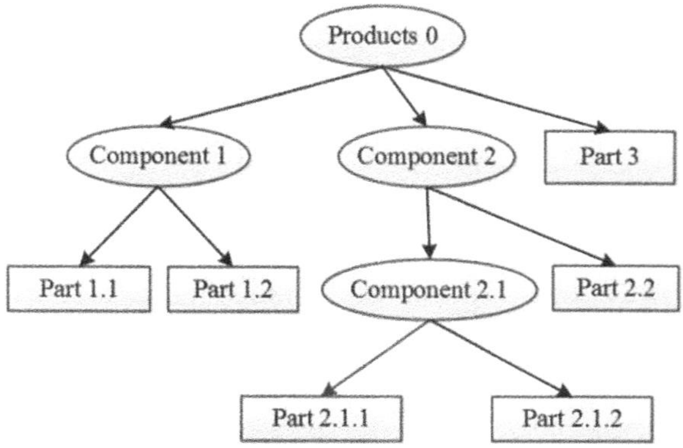

Fig. 5.5 Schematic diagram of the hierarchical model construction method of the assembly model

the information models to form a virtual assembly model of the system. It reasonably describes the information data including the structural relationship between all parts and components in the product assembly and the cooperation constraint relationship. Thus, the assembly of product can achieve the overall abstract expression. Generally, there are two methods for constructing virtual assembly models, based on the hierarchical model (Hierarchical Model) and based on the relation model (Relation Model) [2].

(1) Hierarchical model construction method of assembly model

The schematic diagram of the hierarchical model construction method of the assembly model is shown in Fig. 5.5. The mechanical equipment products are all composed of parts with obvious hierarchical relationships. The product structure formed by this top-down design method has an obvious multi-level style, which is more in line with people's common analytical thinking mode. It can vividly express the structure of the product and can reflect the design thinking of the designer.

(2) Relational model construction method of assembly model

The schematic diagram of the relational model construction method of the assembly model is shown in Fig. 5.6. In the virtual assembly system environment, the assembly constraint relationship is an important component of the assembly information data model. The assembly relationship can constrain the construction of the upper and lower parts of each component. And it can also reflect some of the functions of the product. In addition, the assembly relationship can describe the position and attitude information of components and assembly constraint information.

According to the above analysis of the advantages and disadvantages of the two methods, based on the requirements of the system for the assembly information data model, this section uses a combination of the two methods to construct the assembly information data model of the system, which has achieved clear hierarchical relationship and composition structure of the data model. The assembly relationship data can be expressed comprehensively and accurately, which is convenient for the

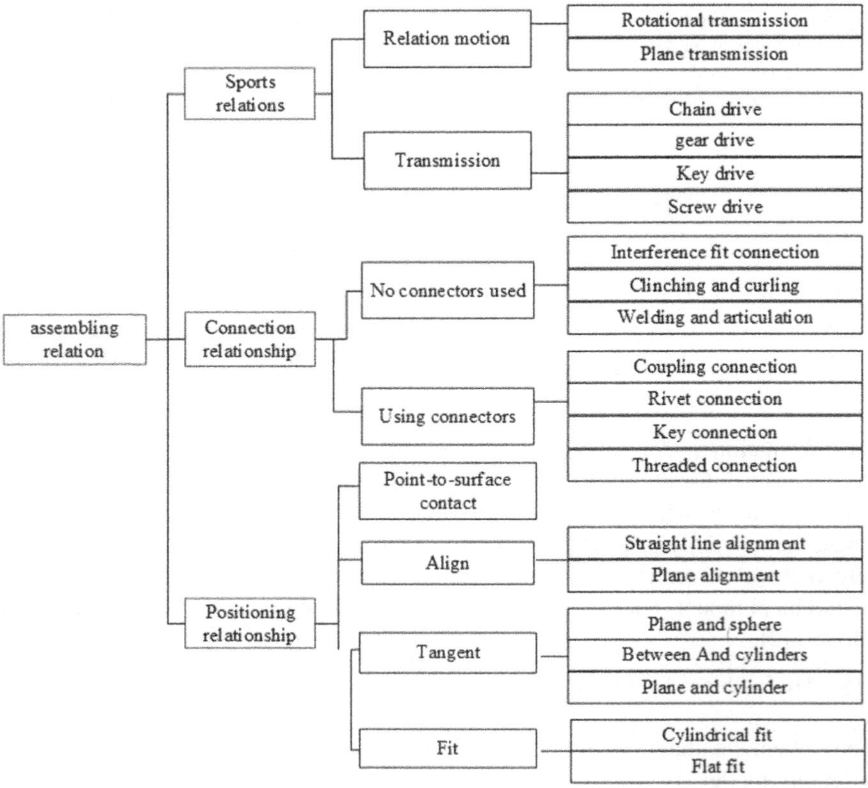

Fig. 5.6 Schematic diagram of the relational model construction method of the assembly model

smooth development of the functional modules of each part of the system and the fluency of the system operation.

5.3.2 *Virtual Assembly Information Framework Based on XML*

In the design of this virtual assembly system, the assembly information should be read from the equipment product CAD assembly model, and the interactive transmission between various functional modules should reduce the information redundancy in the virtual assembly system, and avoid problems such as data inconsistency between various functional modules [3]. In the traditional way of assembly information and data interaction, each functional module is designed according to their own separate standards, which lacks versatility and scalability, and is not conducive to shortening the product assembly design cycle and reducing its design cost [4].

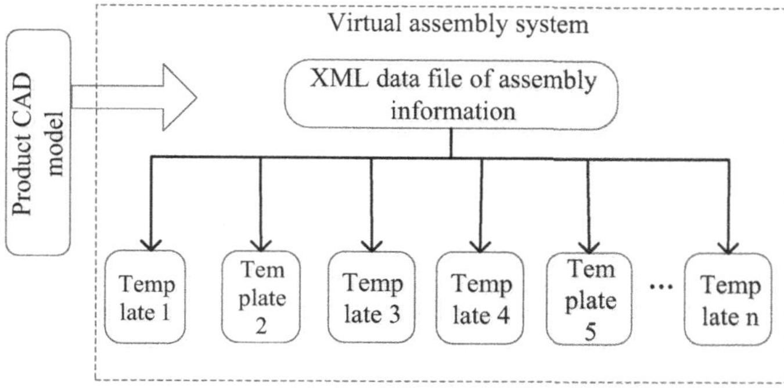

Fig. 5.7 The assembly information data frame of the virtual assembly system of the shearer based on UG

In response to the problems discussed above, this chapter proposes a data interaction transfer method based on XML technology to realize the assembly information data interaction in the virtual assembly system. Based on the UG-based virtual assembly system of the shearer, a data processing framework with a relatively high degree of openness is constructed. XML language [5–7] has many good characteristics, such as complete structure, clear hierarchy, easy to expand, and quickly and efficiently transfer and share information and data between different functional modules [8–10]. Therefore, this scheme can effectively compress the assembly design cycle of the shearer product and improve its assembly design efficiency.

The system assembly information data interaction framework structure is shown in Fig. 5.7. This program uses the XML data file with assembly information data as an intermediate transition module, which can realize the interactive transmission between assembly information data and various functional modules in the virtual assembly system.

5.3.3 Assembly Information Data Model

In view of the clear hierarchical structure of XML grammar, this chapter adopts a hierarchical structure to express the assembly information data of the shearer product. The construction diagram of the hierarchical data model is shown in Fig. 5.8. The first layer is the product layer, which contains the equipment product assembly model file information. The second layer is the component layer, which contains a series of data such as geometric topology information between all the parts that constitute the product model of the equipment and the existing sub-assembly components. The third layer is the assembly data layer, which contains information and data such as the number of fits applied to the corresponding parts and the loading status of the parts. The fourth layer is the assembly coordination constraint information layer. This layer stores all the assembly information data of each coordination, including

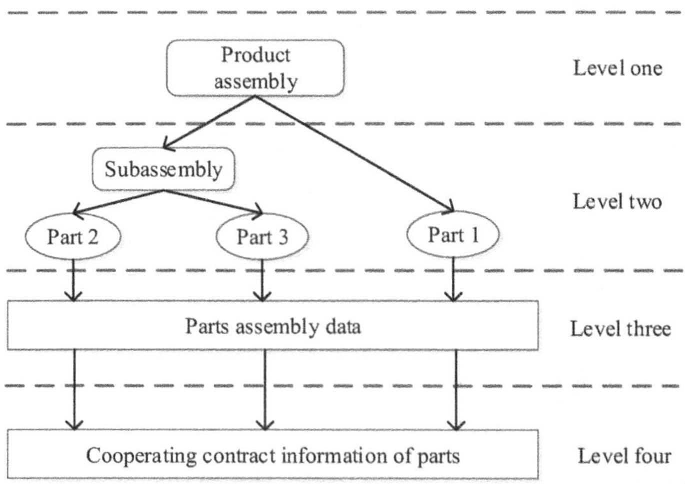

Fig. 5.8 Schematic diagram of the hierarchical structure of assembly information data

the type of coordination, active and passive components of the coordination, the case characteristics of active and passive components, and the type of coordination features.

5.3.4 Assembly Information Data File Based on XML Schema Standard

Unlike traditional database tools such as Access and SQL Server, the XML data format is extremely simple, providing designers with a clear and concise design environment. The grammatical structure adopted by XML is as follows:

- Every starting label must have an ending label corresponding to it.
- In order to simplify the grammar, you can also put the start tag and the end tag in the same label, for example, <active component feature type value = “”/>. The XML parser can translate it into <active component feature type value = “”></active component feature type>. This simplified grammar is only suitable for no further labels in the label.
- The tags must be nested in a proper order, and the start tag and the end tag must be matched strictly in the order of mirroring.
- All parameters in the label must be assigned values.
- All parameter values must be enclosed in double quotation marks.

5.3.5 Construction of Assembly Information Model

This system can analyze and solve the product assembly model, and then output the product assembly information data, and generate the assembly information data document based on XML Schema. Through this document, the information and data interactive communication between the modules is realized in the system, including the information data integration function, assembly information display function, automatic assembly function, assembly planning function, assembly simulation function and other modules.

(1) Establishment of multi-level assembly information model

In the process of constructing the assembly information data model document, the system must complete the import of the product CAD model, the solution of the assembly information data in the product assembly, and the compilation and storage of the assembly information data document. In this section, the UG/Open API secondary development tool is used to implement the relevant functions that need to be cited in the above operation steps and the secondary development of the information window tool in UG is applied to the compilation and storage of assembly information data files. Some UF functions that need to be applied and their uses are shown in Table 5.1.

Open the CAD model of the product assembly through the function UF_UI_open_part (). Use the function UF_ASSEM_ask_component_data() to solve the storage path and name of each component in the product model. Then apply the function UF_ASSEM_ask_inst_of_part_occ () to solve the identification of each part. Then call the function UF_ASSEM_ask_mc_data_of_compnt () to solve the assembly information of each component. Finally, use the function UF_UI_write_listing_window () to output it in the information window according to the XML Schema syntax structure, and use the function UF_UI_save_listing_window () to save the data in the information window as a "*.xml" document. The specific steps are as follows.

Table 5.1 Related UF functions used to build assembly information data model

UF function	Function and purpose
UF_UI_open_part ()	Display the "Open Part" dialog box, and then open a part and set it as a working part. This chapter is used to import assembly CAD models
UF_ASSEM_ask_root_part_occ ()	Obtain the case ID of the assembly root node according to the component ID. This chapter is used to obtain the root node of the product assembly model
UF_ASSEM_ask_part_occ_children ()	Get the component case identifier array and the number of cases under the specified component case. This chapter is used to obtain the component identification and number of sub-assembly components under the corresponding assembly

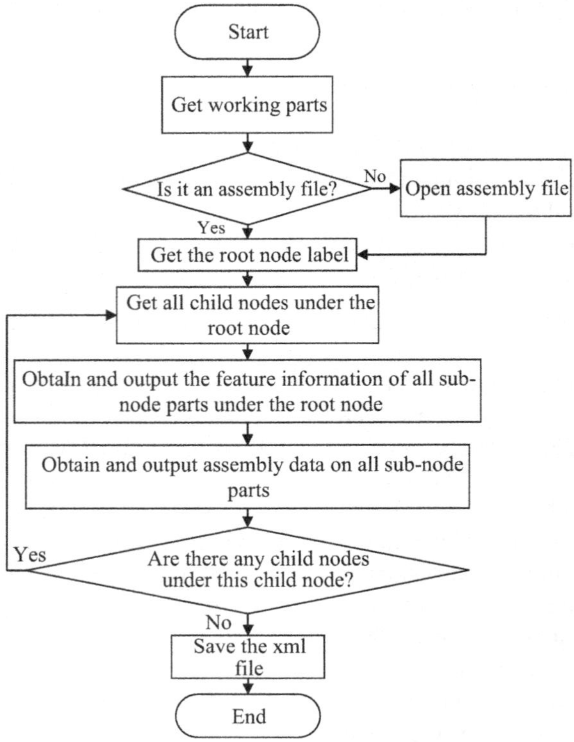

Fig. 5.9 Algorithm flow chart for constructing system assembly information data model

Step 1: Open the information window and output the beginning format of the XML file;

Step 2: Search for the identification of each part of the product assembly model and store it in the array child_part_occs [n], where n is the number of parts in the product)

Step 3: for (int i=0; i<n; i++)

{

Solve the assembly information of child_part_occs [i] and output it to the information window according to the XML Schema syntax.

}

Step 4: Save the data in the information window as a "*.xml" document;

Step 5: The construction of the data file is over.

5.3.6 *Algorithm for Establishing Assembly Information Model*

Figure 5.9 shows the flow chart of the program algorithm for constructing the system assembly information data model document in this chapter. When the program starts, it is necessary to judge the system environment and check the working parts in the current environment.

- If the result is that there are no parts in the system or there are working parts in the system but not assembly parts, the system prompts "the assembly parts that are not working in the system or the working parts are not assemblies, click OK to open the assembly parts".
- If the resultant work part is an assembly, it will traverse analysis to solve all the sub-nodes under the assembly, and then judge whether the parts under the sub-node are sub-assembly parts. If yes, continue to solve and analyze all the sub-nodes of the sub-assembly component in the above-mentioned manner. If it is not a sub-assembly part, but a single part, the execution ends the current program and jumps out of the loop.
- In this way, the recursive algorithm is used to traverse all the parts of the product assembly, and the attached assembly information data is solved, and then output to the text file according to the XML format. When the solution is completed, the text file with the assembly information data is saved as a "*.xml" file, and the algorithm ends.

5.4 The Function Realization of the Virtual Assembly System of the Shearer Based on UG

5.4.1 Key Data Planning of Virtual Assembly Environment

In terms of data processing, a fully functional virtual assembly system should have the following characteristics. First of all, a model with reasonable structure and easy to transmit assembly information data is needed. It is also necessary to reasonably plan the basic attributes and key data information of each product component in the virtual assembly environment. Basic attributes include static data and dynamic data. Static data includes geometric features, pose status, and so on. The dynamic data includes the movement and rotation vectors of the motion in the virtual assembly system environment, and the corresponding process steps (Process Step), etc. The key data information refers to the dynamic planning data of each component in the operating environment during the operation of the system. These features provide an effective, reliable and complete data foundation for the system to realize functions such as automatic assembly, assembly sequence, and assembly path planning and assembly process simulation.

(1) Overview of key data in virtual assembly environment

This section needs to study the relevant data of product parts and assemblies in the UG assembly environment.

The coordinate system is also a vital data in the UG assembly environment. The coordinate system matrix and the transformation matrix determine the relative position of the components in the assembly environment. The input and output parameters of some functions related to assembly are represented by coordinate system matrix and transformation matrix. In UG/Open API, commonly used terms related to the coordinate system are space (space) and assembly space (assembly space). Space (space) refers to the space where the component is located when it is created, that is, the space corresponding to its absolute coordinate system. Assembly space (assembly space) refers to the space in which components are loaded into the assembly environment.

The transformation matrix used to describe the component from space to the current assembly space is a 4 × 4 real number matrix as shown below.

$$\begin{bmatrix} r[0][0] = c[0], r[0][1] = c[3], r[0][2] = c[6], r[0][3] = o[0] \\ r[1][0] = c[1], r[1][1] = c[4], r[1][2] = c[7], r[1][3] = o[1] \\ r[2][0] = c[2], r[2][1] = c[5], r[2][2] = c[8], r[2][3] = o[2] \\ r[3][0] = \quad 0, r[3][1] = \quad 0, r[3][2] = \quad 0, r[3][3] = \quad 1 \end{bmatrix}$$

The three-dimensional coordinates of the origin of the part are stored in three elements: r[0][3], r[1][3] and r[2][3], where: r[0][3] = o[0] is The x coordinate of the origin, r[1][3] = o[1] is the y coordinate of the origin, and r[2][3] = o[2] is the z coordinate of the origin.

In the coordinate system matrix (r[0][0], r[1][0], r[2][0]), (r[0][1], r[1][1], r[2][1]) and (r[0][2], r[1][2], r[2][2]) are three column vectors, where:

c[0], c[1], c[2] are the values of i, j, and k in the coordinate X-axis direction vector (i, j, k);

c[3], c[4], c[5] are the values of i, j, and k in the coordinate Y-axis direction vector (i, j, k);

c[6], c[7], c[8] are the values of i, j, and k in the coordinate Z-axis direction vector (i, j, k);

r[3][0], r[3][1] and r[3][3] are not used, r[3][3] represents the conversion ratio, and the value is always 1.

(2) Key data planning method of virtual assembly environment

When assembling the key information planning, it is mainly for the relevant function modules in the system to make targeted selections. During assembly planning, each component includes an assembly/disassembly action, which makes the necessary planning for the parameters in Table 5.2 when planning key data.

Table 5.2 Planning parameter list

Parameters	Annotations	Parameters	Annotations
(a) Global parameters			
Ug_Matrix [4] [4]	Pose matrix	Ug_Vector [3]	vector
Ug_Point3D [3]	Point		
(b) *Assembly model(VA_component)*			
instance_tag	Instance ID	occurence_tag	Case ID
part_name	Part name	instance_name	Instance name
refset_name	The name of the reference set	ini_pos	The initial pose matrix of the component
pos	Component pose matrix	children	First subcomponent
next	The next part of the same level	parent	Parent part
type	Assembly parts	unit	unit
key	Judgement parameters for disassembly execution		
(c) Component movement(VA_movement)			
Translation	Translation vector	point	Point on the axis of rotation
rotation	Rotation axis direction	angle	Rotation angle
prev	Previous move	next	Next move
comp	Move the corresponding parts		
(d) All moves(VA_movements)			
head	First mobile node	tail	Last mobile node
move_num	Number of moves		
(e) Assembly(VA_assembly)			
head_comp	Assembly root node	moves	Moving pointer
file_name	File name of the assembly		
(f) Part load position data(VA_component_load_location_data)			

(continued)

Table 5.2 (continued)

Parameters	Annotations	Parameters	Annotations
parent	Parent part	prev	Previous sibling part

Note The above parameters are quoted in the source program of the application program written to realize this function

5.4.2 Automatic Assembly

For the automatic assembly technology, the basic process is: establishment of all part models → manual assembly in the CAD software assembly environment → read product assembly information (including all product parts information, assembly coordination relationship data, etc.) → carry out secondary development of the software → Develop the user interface of the application → Run the program to realize the automatic assembly of the product model. In addition, the parameter relationship between the components established in the assembly of the product model can be automatically generated after the product model is updated by changing the relevant parameters of certain components. This mode has parameterization and automatic assembly meaning. The realization of automatic assembly function can effectively improve the efficiency of assembly design, reduce manual operation and reduce the difficulty of design.

(1) Assembling related technologies in the secondary development of UG

To realize the automatic assembly of product parts models is essentially to perform corresponding operations on the matching relationships between parts in the assembly environment. In the assembly environment of UG/Open API, each component in the assembly file needs to be positioned by constraining the mating relationship with other parts.

The establishment process of each assembly constraint is roughly divided into five steps. The first step, data definition, is mainly to complete the definition of the data structure of the assembly constraint information. The second step is to initialize the cooperation relationship, and call the function UF_ASSEM_init_mc() to initialize the assembly relationship defined in the first step. The third step is to solve the assembly constraints and call the function UF_ASSEM_solve_mc() to solve the above coordination relations. The fourth step is to apply assembly constraints, call the function UF_ASSEM_apply_mc_data() to apply the solved assembly constraints to the corresponding parts. The fifth part is to update the model, call the function UF_MODL_update() to update the product model and complete the automatic assembly.

(2) Description of automatic assembly algorithm

To realize the automatic assembly function of the virtual assembly system, it is necessary to read and call the part assembly information data in the XML document containing the product model assembly information data, and then apply it to the corresponding parts to complete the entire product assembly model Automatic assembly.

The automatic assembly of the parts of the entire product assembly model needs to use the function UF_ASSEM_add_part_to_assembly() to load all the parts in the product model into the assembly environment one by one according to the assembly information data of the product XML document. According to the assembly coordination data in the XML document, use functions such as UF_ASSEM_solve_mc() to automate the assembly of the product model until the assembly and coordination relationships of all components are fully applied.

5.4.3 Assembly Sequence and Assembly Path Planning

The planning technology of assembly sequence and assembly path is a crucial link in the development of virtual assembly system. Usually people call this process as assembly sequence planning ASP (Assembly Sequence Planning) [11]. The main process of assembly sequence planning is the disassembly and assembly planning of the assembly model based on the knowledge of the product production process in the virtual assembly environment to generate a reasonable assembly sequence and assembly path, so as to make the assembly plan feasible, practical and low cost [12].

5.4.4 Dynamic Simulation of Assembly Process

(1) Representation and transformation of component pose in assembly space

In the UG assembly environment, the pose (position and posture) state of the assembly components is represented by a "4*4" pose matrix [P], in the form as follows:

$$[\mathrm{P}] = \begin{bmatrix} X_{V1} & Y_{V1} & Z_{V1} & X_S \\ X_{V2} & Y_{V2} & Z_{V2} & Y_S \\ X_{V3} & Y_{V3} & Z_{V3} & Z_S \\ 0 & 0 & 0 & 1 \end{bmatrix} \tag{5.1}$$

Among them (XV1, YV1, ZV1), (XV2, YV2, ZV2), (XV3, YV3, ZV3) are the relative coordinate systems of the assembly parts in the assembly space, used to represent the posture of the assembly parts, (XS, YS, ZS) is used to represent the

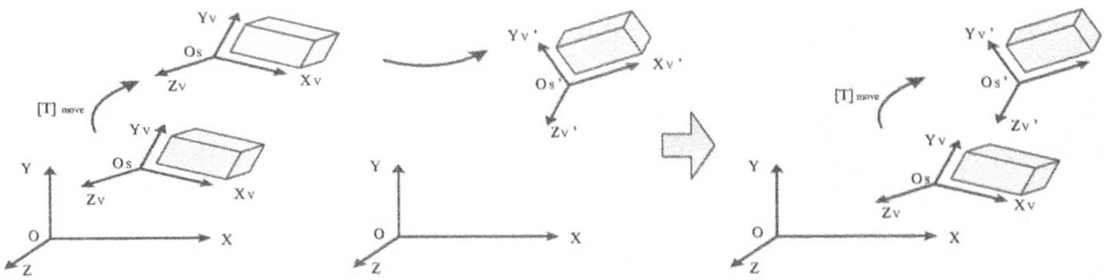

Fig. 5.10 Description of the implementation scheme of pose transformation. **a** Schematic diagram of position transformation. **b** Schematic diagram of posture transformation. **c** Schematic diagram of posture composite transformation

position coordinates of the assembly component in the assembly space. The first three elements in the last line are not used, and the last "1" represents the ratio.

Figure 5.10 shows the schematic diagram of the realization scheme of the pose transformation of assembly parts in the virtual assembly system. Figure 5.10a shows the position transformation of assembly parts. Figure 5.10b shows the posture transformation of the assembled parts. The position transformation shown in Fig. 5.10a and the posture transformation shown in Fig. 5.10b are superimposed to realize the composite transformation of the assembly parts shown in Fig. 5.10c. Therefore, the complete pose transformation of each component assembly is realized by compound superposition of its position transformation and posture transformation.

The transformation of the pose state of the assembly parts in the virtual assembly environment is essentially the transformation of the assembly parts from the original pose matrix to a new pose matrix. The realization method is to transform the original pose matrix [P0] of the assembled parts into the final pose matrix [Pt]. This process requires the aid of a "4*4" pose conversion matrix [T]. The conversion operation is as follows:

$$[\mathrm{Pt}] = [\mathrm{T}] \cdot [\mathrm{P0}] \tag{5.2}$$

The format of the pose transformation matrix [T] is as follows:

$$[T] = \begin{bmatrix} X_1 \; Y_1 \; Z_1 \; M_{VX} \\ X_2 \; Y_2 \; Z_2 \; M_{VY} \\ X_3 \; Y_3 \; Z_3 \; M_{VZ} \\ 0 \quad 0 \quad 0 \quad 1 \end{bmatrix} \tag{5.3}$$

The matrix (MVX, MVY, MVZ) represents the moving components of the component position transformation in the three coordinate axis directions of the assembly space. (X_1, Y_1, Z_1), (X_2, Y_2, Z_2), (X_3, Y_3, Z_3) respectively represent the relative coordinate system transformation of the assembly component in the assembly space coordinate system to realize the transformation of the component posture. The meaning of the elements in the last row is the same as in the pose matrix.

5.5 Development of Coal Shearer Virtual Assembly System Based on UG

5.5.1 System Application Framework

Before the system function modules are developed and implemented, the system framework must be designed first, and the program interface elements composed of the system framework must be designed and created, such as dialog boxes, menus, toolbars, and function files in the system. Then it is necessary to study the related technologies in the open application program interface of UG/Open API. The main content of UG/Open includes [11] user interface (dialog window) design (UI Styler), menu script language (Menu Script), application programming interface (API), graphical interactive programming (GRIP).

Among them, application programming interface technology (API), user interface (dialog window) design (UI Styler) and menu script language (Menu Script) are the main contents that need to be studied in this chapter to develop a virtual assembly system for coal mining machinery.

(1) System application program composition
In UG/Open API, there are two ways to stimulate the application by menu. One is "the menu directly stimulates the application"; the other is "the menu stimulates the dialog box", through the corresponding program of the dialog box [12]. Because the program structure is simple and straightforward, the former method is compatible with the software environment. But its program is concealed, which is not conducive to manual intervention, and lacks the convenience of human–computer interaction. The latter can solve the above-mentioned problems. However, due to its complicated program structure and no direct connection to the software environment, its compatibility is relatively poor compared with the former. Since these two methods have their own advantages and disadvantages, the system will adopt different menu excitation methods according to the specific requirements of the program.

(2) System Directory Structure
UG secondary development applications need to create a directory structure in compliance with UG/Open rules under the project path. When the UG software starts, it will automatically load the applications and resources in the above-mentioned directory [13]. There are many such directories, the most commonly used are "Startup", "Application" and "Document". The directory structure is shown in Fig. 5.11.

Fig. 5.11 System directory structure

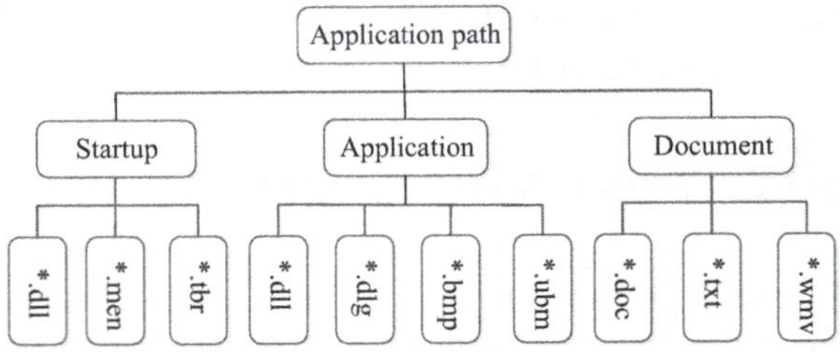

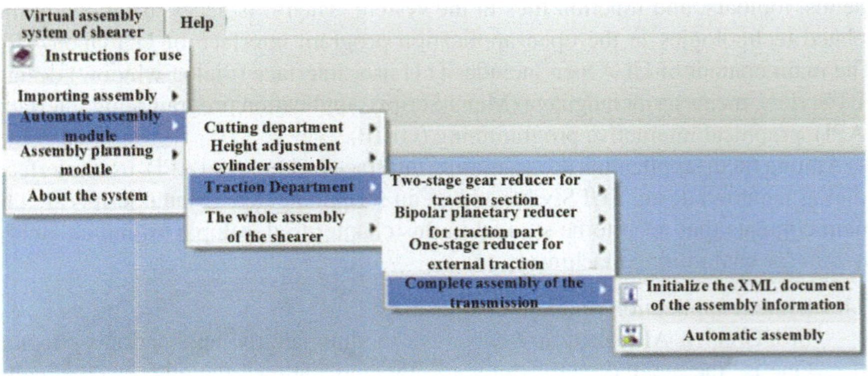

Fig. 5.12 The virtual assembly system menu of the shearer based on UG

5.5.2 System Main Menu Design

UG/Open Manuscript is used to create UG tailoring script, its extension is "*.men" [14]. Design the system menu files separately, and then put them into the "Startup" folder of the system directory. The software interface is shown in Fig. 5.12.

5.5.3 Assembly Import Module Development

The function of the assembly import module is single, but the structure is complex. Therefore, the module focuses on the design of the menu structure when the program framework is established. The whole structure of the shearer is divided into several sub-assembly parts for separate assembly and import planning. The menu structure of this function module is shown in Fig. 5.13. The most important UF function is UF_ASSEM_add_part_to_assembly ().

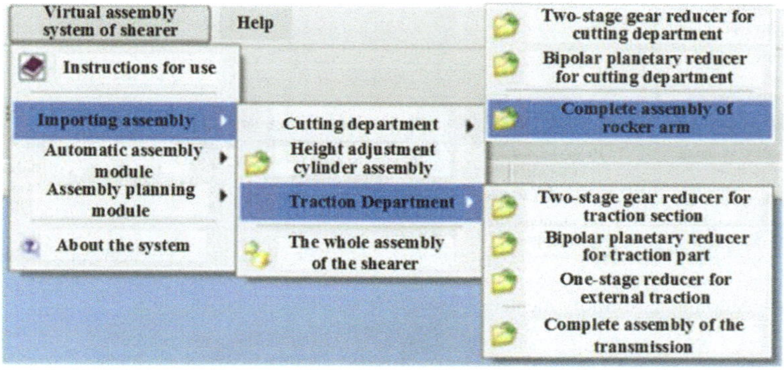

Fig. 5.13 Model import menu

5.5.4 Automatic Assembly Module Development

(1) Dialog interface design

In the application software developed by UG/Open API, there are two types of dialog boxes. One type is the UG style dialog box provided in UG/Open. The other is the Windows style dialog box established in the Microsoft Visual Studio 2010 programming tool.

During the development of the automatic assembly module, it is necessary to display the assembly information such as the cooperation relationship and the assembly structure tree. Because of the specific requirements for the control, the Windows style dialog box is selected to establish the dialog box window of the automatic assembly module.

(2) Establishment of procedural framework

The execution of the program of the automatic assembly module needs to activate the above-mentioned dialog box from the button of the menu, and the dialog box executes the corresponding function. The menu design of this function module is the same as the menu structure of the model import module. The difference is that selecting the menu triggers the dialog mode.

The realization of the related functions of the automatic assembly module is inseparable from the data support of the XML assembly information data file. The menu design is shown in Fig. 5.14. In the first level menu, select "Automatic assembly module" → "Cut part" → "Cut part two-level spur gear reducer" to open the fourth level menu, and click "Assembly information XML document Update" can update the assembly information of the cutting department's secondary spur gear reducer. The XML file is convenient for the system to call. Click "Auto Assembly" to open the automatic assembly dialog box of the cutting department's secondary spur gear reducer, and then you can execute the corresponding Function.

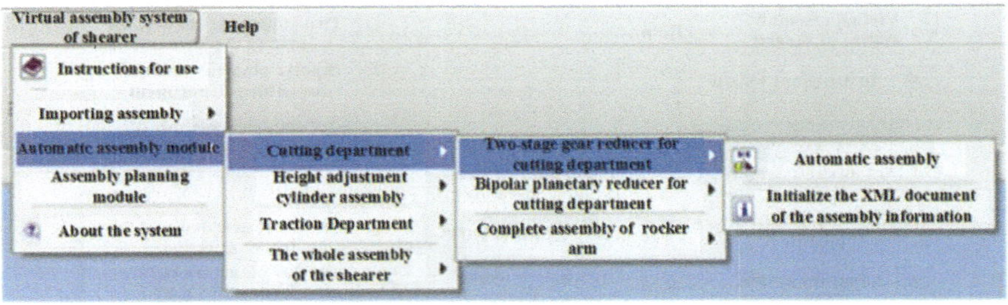

Fig. 5.14 Automatic assembly menu

5.5.5 Development of Assembly Planning Module

After the parts of the corresponding model are imported into the virtual assembly system and the automatic assembly is completed, the assembly planning of the assembly parts can be carried out, including functions such as assembly sequence, assembly path, and dynamic simulation of the assembly process.

(1) Dialog interface design

There are many related functions in the assembly planning module of the system, and it is necessary to establish a dialog box to integrate all the function buttons. Because these functions do not need other types of controls to run in the background, the UG style dialog box is selected to design the dialog box interface of the assembly planning module.

Click the control in the "Control Bar" to add the corresponding control to the dialog box, and at the same time, the indicator of the control will be displayed in the "Object Browser". Select the control in the "Object Browser" to edit the selected control in the "Resource Editor". After designing the dialog box, save the file, save the file as "*.dig", and select "C language" as the language. Three files "*.dig", "*.h" and "*.c" will be generated after saving.

(2) Procedure framework establishment

The assembly planning module is the same as the automatic assembly module. Use the dialog box activated by a menu to activate related function applications. Write application programs in accordance with the relevant functional program flow designed in this chapter. Put the "*.dll" file of the application program into the "Startup" folder of the system directory to build the system framework. The assembly model needs to be initialized before the assembly is planned. If the assembly plan is not initialized, a dialog box will pop up to prompt "Please initialize the assembly model first!". After initializing the model, click "Assembly Planning" to pop up the dialog box shown in Fig. 5.15, which allows you to dynamically plan the assembly model.

(3) Example application of assembly planning

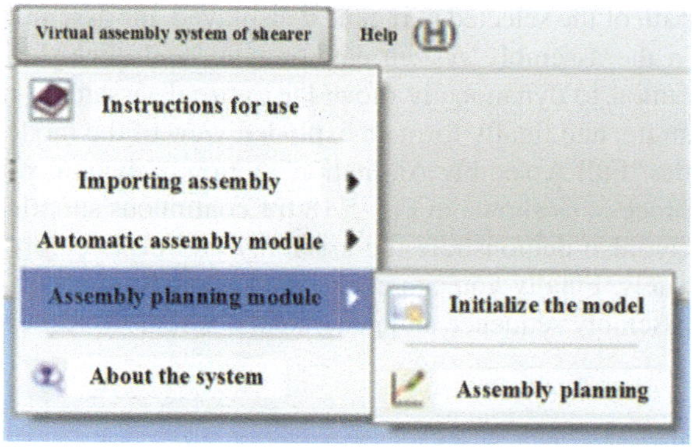

Fig. 5.15 Assembly planning dialog box

According to the program prompts, after completing the assembly planning of the planetary gear structure of the shearer cutting part, the dialog box is shown in Fig. 5.16 to display the planned parts list. This module also saves planning data, reads planning data, and restarts Planning function. After the planning is completed, the assembly

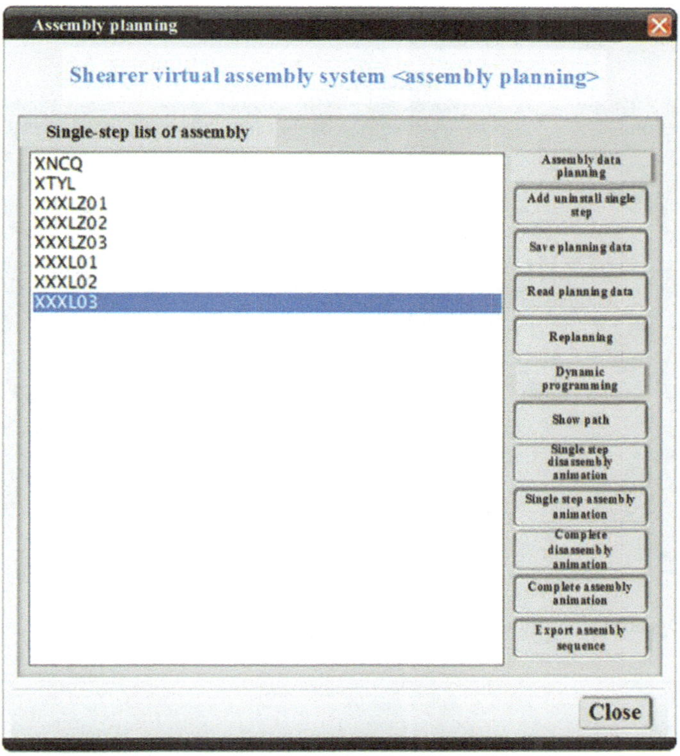

Fig. 5.16 Assembly planning interface

path of the selected part can be displayed, the assembly path is highlighted in yellow in the assembly system environment, and click the "full disassembly animation" button to dynamically move the entire disassembly process in the system Demonstrate, and finally form an exploded view of the model, as shown in Fig. 5.17. Click the "Full Assembly Animation" button to dynamically display the entire assembly process, as shown in Fig. 5.18 for continuous shooting of assembly animation. You can also demonstrate the assembly and disassembly animation of each part separately. Finally you can click the "Export Assembly Sequence" button to export the assembly sequence and generate the assembly sequence file.

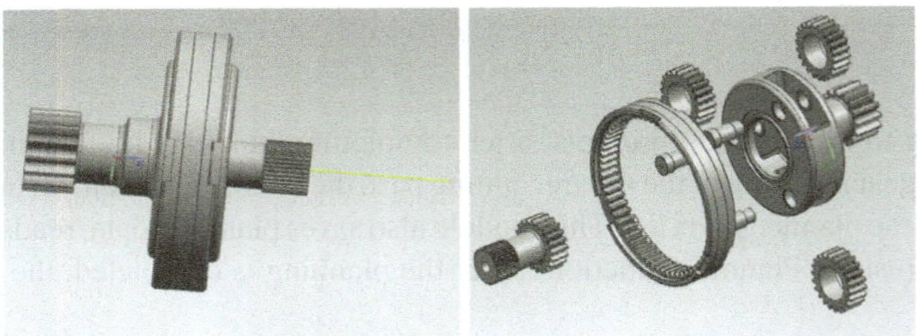

Fig. 5.17 Assembly path display and exploded view

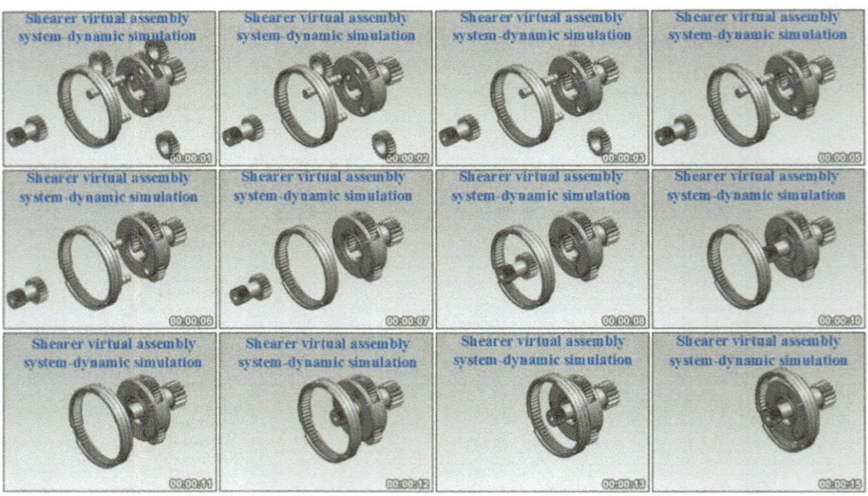

Fig. 5.18 Dynamic simulation of assembly process

References

1. Hou, Y., Ding, X.: UG/Open Secondary Development Examples. Chemical Industry Press, Beijing (2007)
2. Sheng, Y.-j, Zhang, Z.-m, Fan, Q.-j, et al.: On the CAPP system for aircraft assembly based on CAPP framework. Mech. Sci. Technol. **1**, 145–146 (2001)
3. Zhang, M., Xu, Q., Shen, X.: Data exchange and sharing platform model based on XML. J. Tsinghua Univ: Nat. Sci. Ed. **43**(1), 105–107 (2003)
4. Hu, W., Wei, X.: XML-based process planning information modeling and exchange. Modul. Mach. Tool Autom. Process. Technol. **12**, 24–26 (2005)
5. Chen, S.: XML and it's application. Appl. Res. Comput. **19**(3), 115–117 (2002)
6. Sato, Y., Morishige, K.: Development of operation planning support system using XML database of machining information. J. Jpn. Soc. Precis. Eng. **74**(1), 92–96 (2008)
7. Šormaz, D.N., Arumugam, J., Harihara, R.S., et al.: Integration of product design, process planning, scheduling, and FMS control using XML data representation. Robot. Comput.-Integr. Manuf. **26**(6), 583–595 (2010)
8. Cui, Z.-h, Wang, J.: Research on two-way system integration between PDM and 3D CAD based on XML. Mach. Des. Manuf. **2**, 79–81 (2012)
9. Fang, W., Wu, H.: A comparative analysis of XML DTD and schema—DC as example. Inform Sciences. **22**(4), 506–509 (2004)
10. Shen, L.-q, Yang, J., Zhang, L.-p: Research on achieving mechanism and methods of automatic assembly in CAD system. Mach. Tools Hydraul. **11**, 64–68 (2006)
11. Zhu, J.-g: Virtual assembly key technologies and its application in customized furniture products. J. Nanjing Forestry Univ.: Nat. Sci. Ed. **37**(2), 129–135 (2013)
12. Huang, X., Li, Y.-g: UG Application Development Tutorials and Detailed Examples. Tsinghua University Press, Beijing (2005)
13. Zhou, L.-z, Li, Q.-z, Qin, K.: Secondary Development Based on UG NX System. Jiangsu University Press, Zhenjiang (2012)
14. Dong, Z., Tian, L., Fu, Y.: UG/Open API Programming Foundation. Tsinghua University Press, Beijing (2002)

Chapter 6
Human–Machine Interaction Method of Virtual Assembly

6.1 Introduction

Virtual assembly human–computer interaction is a kind of interactive action matching with the characteristics of Three-dimensional environment itself, which enables users to get a direct immersive feeling in the virtual scene. In the virtual reality assembly system for coal machine equipment, various human–computer interaction devices are used to enable designers and researchers of coal machine equipment to interact with the system with more abundant means, so as to find the problems existing in the equipment and provide strong technical support for equipment improvement and new product research and development.

Three-dimensional interaction technology is a "control-display" mapping relationship, user devices (such as mouse, keyboard, data gloves, force feedback device, etc.) input control information to the system, and then the system output the execution results to the user. There are many tasks involved in 3D interaction, and the main human–machine interaction in this system is assembly interaction. The assembly interaction is achieved through three methods: one is based on the traditional mouse and keyboard through the operator or robot to realize the full-scale browsing of the scene to realize the interaction; the second is based on the data glove and the position tracker to realize the interaction, the last one is a human–computer interaction technology which is based on force feedback. The specific implementations of these interaction modes will be introduced below.

6.2 Mouse and Keyboard Human–Computer Interaction

6.2.1 Scene Interaction Based on Trackball

The assembly scene interaction does not refer to the operation of the entire assembly scene (including the objects in it), but to change the position of the camera or observer

X. Wang et al., *Virtual Reality Technology in Mining Machinery*,
https://doi.org/10.1007/978-981-16-4408-5_6

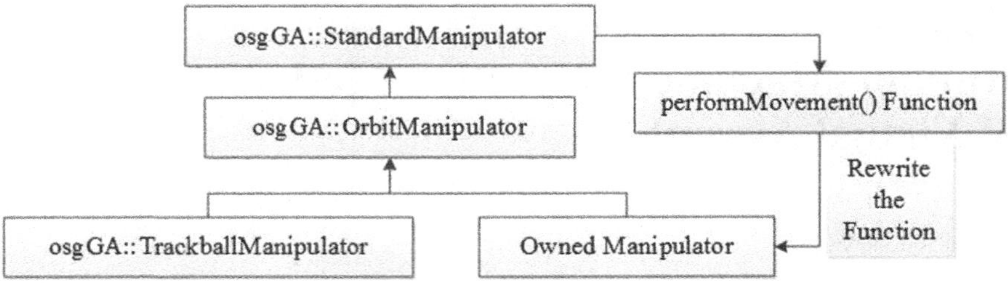

Fig. 6.1 The realization principle of the scene interactive operator

observing the assembly scene to facilitate the operator to observe and understand the scene from different angles.

OSG comes with a trackball scene browser OSGGA::TrackballManipulator, which is an operator based on the trackball algorithm, which can better realize the browsing of the scene. The left button is mainly responsible for the rotation operation of the scene, but in the assembly process of dragging the object on the coordinate axis, the left button is often required to select the assembly model, in order to not only have the advantages of the trackball manipulator, but also enable the left mouse button to assemble For the operation of the model, this chapter rewrites the operator to meet the functional requirements of the system. By looking at the OSG source code, it is found that the trackball manipulator is inherited from OSGGA::OrbitManipulator, so this chapter also inherits from it to get the required manipulator. Further research can find that OSGGA::OrbitManipulator in OSG is inherited from OSGGA::StandardManipulator, and the performMovement() function in OSGGA::StandardManipulator is responsible for assigning the response function of the mouse event, so the original mouse is changed by rewriting this function Responding to events, as shown in Fig. 6.1.

6.2.2 *Mouse-Based Assembly Interaction*

Assembly interaction uses interactive events to translate, zoom, and rotate the selected model without making any changes to the observer. In assembly interaction, a Matrix-Transform parent node is generally set for the model to be operated, and the model is operated by changing the transformation matrix of the parent node. This is also the implementation of the drag and drop device in the OSGManipulator library.

For the mouse to achieve assembly interaction, there are mainly two methods. One is to move and rotate the model on the coordinate axis based on the coordinate axis dragger. This method is intuitive to operate and has high accuracy; the other is to directly drag the mouse. The moving model is moved on a two-dimensional screen, and then mapped to a three-dimensional space to realize the movement of the model. This method is more simple and convenient to operate, but it is not intuitive, requires strong spatial imagination, and is not accurate.

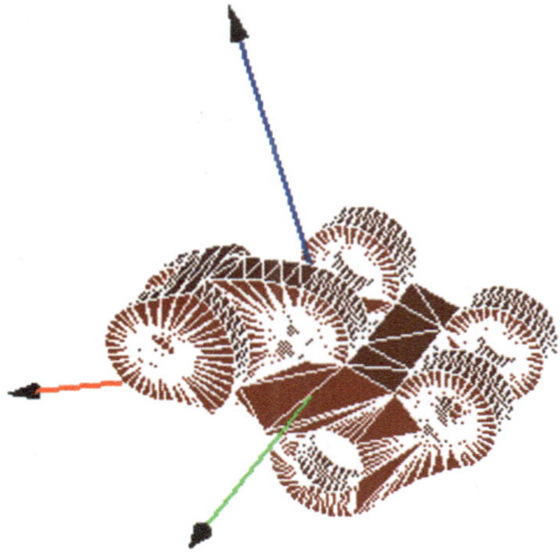

Fig. 6.2 The effect diagram of the cutting part oil bracket being selected

Based on the above analysis, this chapter uses the coordinate axis dragger to realize the interaction of the mouse to the assembly scene, which mainly needs to realize the following two points:

- Press the left mouse button to select the assembly model and highlight it.
- Drag the coordinate axis dragger with the mouse to drag the assembly model.

Figure 6.2 is an effect diagram of selecting the oil bracket when assembling the cutting part of the shearer.

After selecting the model, you need to use the drag and drop tool to move the model. The principle is to first trigger the intersection of the ray and the coordinate axis by clicking the mouse to obtain the specific selected axis system, and then map the movement of the mouse on the two-dimensional screen to the coordinate system to realize the dragger in its own coordinates. The movement of the system, and then through the CommandManager to transfer the operation of the drag to the candidate object (Selection), and the candidate object manipulates the model in the scene for the final movement. The working principle of the drag and drop is shown in Fig. 6.3.

6.3 Virtual Hand Human–Computer Interaction Subsystem

A virtual hand is a hand simulated in a virtual environment, which provides a new means of interaction between the real environment and the virtual environment. In the virtual environment, the interactive operation of the virtual hand is realized through the cooperation of the data glove and the position tracker [1].

Fig. 6.3 Working principle of Dragger

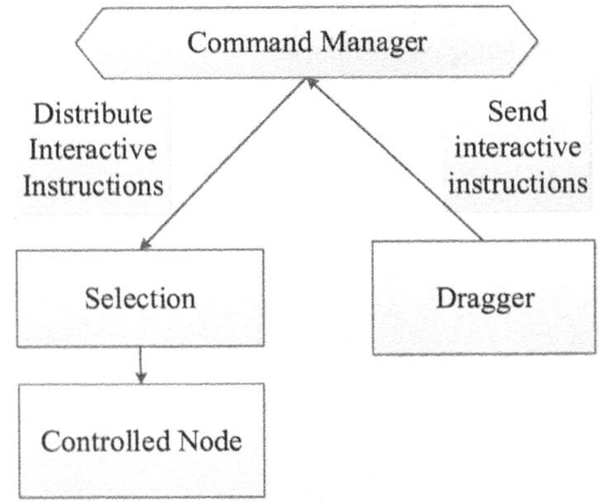

6.3.1 Technical Route

This system uses the data glove 5DT Glove and the position tracker Polhemus LIBERTY as the system hardware equipment, establishes the virtual hand model through 3DMAX, and uses the virtual disassembly and assembly part of the OSG/IVE model in the virtual reality resource library established in Chap. 3 as the operating objects. Under the new virtual reality platform, the process of virtual hand

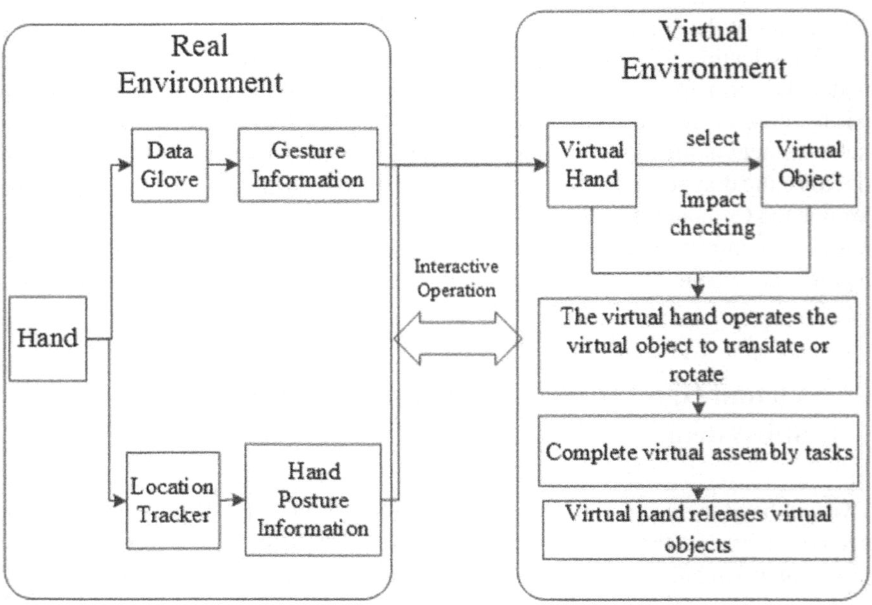

Fig. 6.4 The implementation process of virtual hand human–computer interaction assembly

man–machine interactive assembly is realized. Figure 6.4 shows the implementation process of virtual hand man–machine interactive assembly.

In this system, the movement between the virtual hand and the human hand is adjusted through the data glove. The user can use virtual hands instead of human hands to operate the virtual component model, and get information feedback from the virtual environment, so as to realize the interactive experience between the real environment and the virtual environment.

6.3.2 *Establishment of Virtual Hand Model*

6.3.2.1 Analysis of Human Hand Structure and Movement Characteristics

The establishment of the virtual hand model must ensure the following three conditions:

- The model structure of the virtual hand should be roughly the same as the structure of the human hand.
- The operating habits of virtual hands are consistent with those of human hands.
- The virtual hand should take care of each other with the human hand.

The composition of the human hand is more complicated. Combined with the placement of the sensors of the 5DT data glove, the human hand can be simplified into one palm and five fingers in the end. Among the five fingers, the thumb is set with two phalanges, and the rest of the fingers except the thumb are set. The three knuckles are called the distal phalanges, the middle phalanges, and the proximal phalanges. There are 15 parts in total, as shown on the left of Fig. 6.5.

After understanding the mechanism of virtual hand establishment, 3DMAX is used to model the virtual hand [2], and the OSGExp plug-in is still used to export it into OSG/IVE format. The palm can move and rotate, and when the palm moves or rotates, it drives the fingers to move in the same direction. In the OSG virtual scene, the scene is managed through the scene tree, and the virtual scene is established according to the above palm and finger movement rules. The scene structure diagram of the hand uses the entire palm as the root node to construct the structure tree. It is the first level, the five proximal phalanges are the second level, the five middle phalanges are the third level, and the four distal phalanges are the fourth level. Each phalanx is driven by its upper-level phalanx, and its own movement also drives the next-level phalanx. The final virtual hand structure scene tree is shown in Fig. 6.6.

6.3.2.2 Calibration of Data Gloves

This system uses the 5DT Data Glove 5 Ultra right-hand data glove, which consists of 6 sensors and an interactive box, as shown in Fig. 6.7.

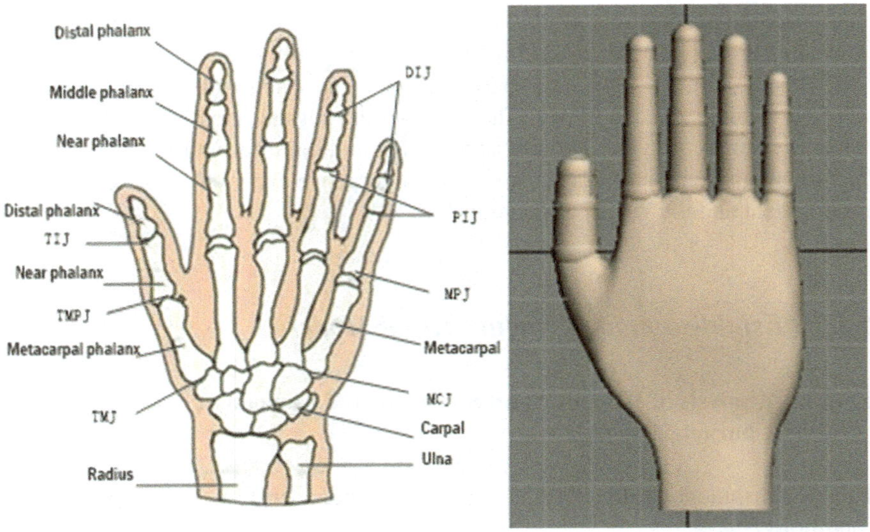

Fig. 6.5 Simplified structure diagram of manpower and virtual hand

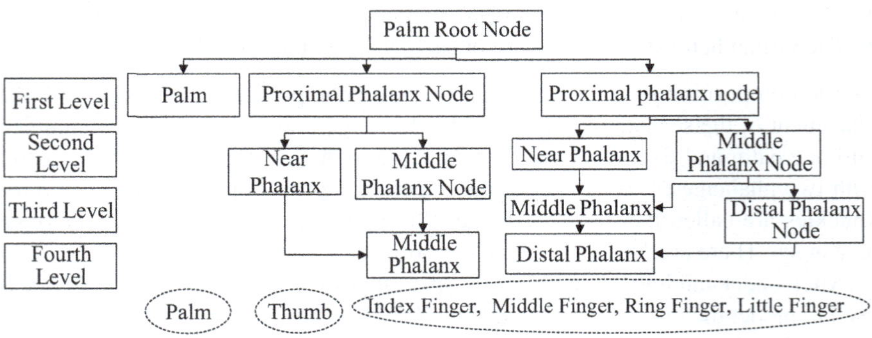

Fig. 6.6 Virtual hand scene structure diagram

Among them, sensor 1 to sensor 5 are curvature sensors, used to test the average degree of bending of each finger; sensors 6 and 7 are tilt angle sensors, used to measure the tilt angle of the palm around two different coordinate axes, according to The real situation of the human hand does not allow 360-degree rotation. Set the tilt angle range from 0° to 255°, and the interactive box is connected to the interface of the graphics workstation through the USB interface [3]. In order to enable everyone to use the data glove smoothly, it needs to be calibrated before use. There are two calibration methods for the 5DT Data Glove 5 Ultra data glove: automatic calibration and software calibration [4].

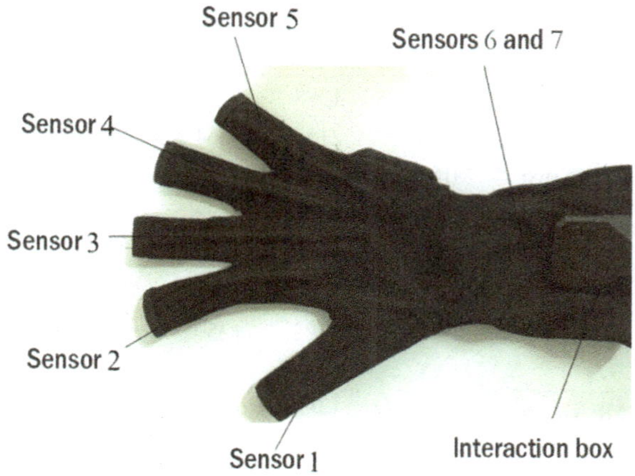

Fig. 6.7 Data glove

Software calibration is to obtain data through the natural movement of five fingers. After multiple movements, you can obtain more satisfactory data. Automatic calibration uses the company's own programming calibration software for testing. Among them, the maximum value is a human hand making a fist. The output value of the sensor during movement, the minimum value is the output value of the sensor when the hand stretches.

With automatic calibration, this method uses the calibration driver that comes with the data glove. After the operator puts on the data glove, the right hand quickly and continuously performs alternate movements of making a fist and stretching as much as possible. The program will automatically record each movement through the sensor. The maximum, minimum and dynamic range, and constantly refresh the data in real time according to the operator's hand movement, so as to be used in calibration calculations.

Here, three different users are selected to calibrate the glove. The data measured by different users using the same action show big differences. Compare the three sets of calibration data and find the average value as the ideal calibration data, as shown in Table 6.1.

Using the automatic calibration method, open the calibration program, put on the data gloves, make a fist and stretch with the right hand in a loop, and each sensor will record the maximum, minimum and dynamic range of each movement, and continuously update it in real time. The test result is shown in Fig. 6.8.

6.3.3 Establishing the Relationship Between Location Tracker and Data Glove

In order for a virtual hand to perform assembly operations in a virtual environment, in addition to the aforementioned data gloves, a position tracker is also needed. The

Table 6.1 Calibration data chart

	Best value	Thumb	Index finger	Middle finger	Ring finger	Little finger
Group One	MIN	1388	1295	1850	1614	1464
	MAX	2155	2664	2834	2800	2800
Group two	MIN	1383	1243	1855	1723	1503
	MAX	1966	2288	2852	3078	2761
Group Three	MIN	1383	1240	1833	1682	1482
	MAX	2274	2423	3029	3110	2786
Average	MIN	1382	1259	1846	1673	1483
	MAX	1834	2458	2905	2996	2782
MEA		1484	1807	2264	2550	2049
F		0.226	0.457	0.395	0.663	0.436

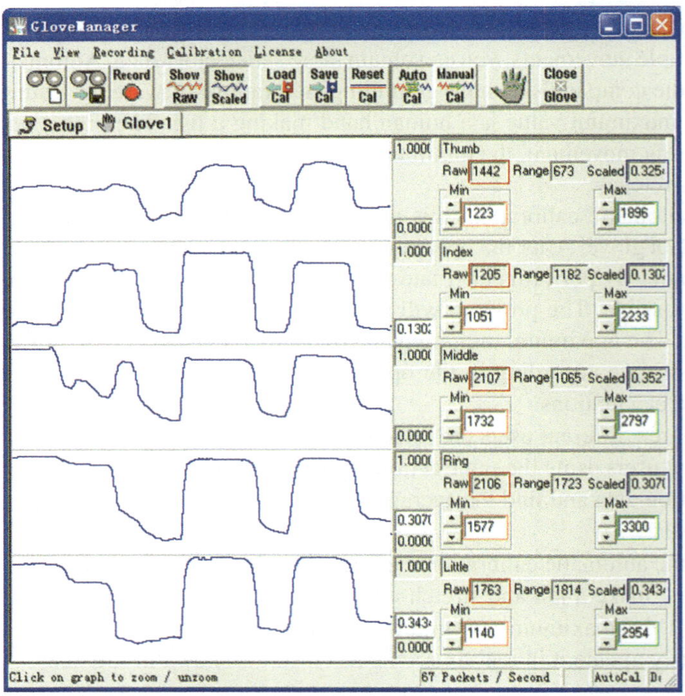

Fig. 6.8 Data glove test chart

position tracker is mainly composed of a position sensor, a fixed coordinate source and a signal converter. The sensor and the fixed coordinate source are connected to the signal converter through the RS232 serial interface, and the signal converter is connected to the graphics workstation through the USB interface.

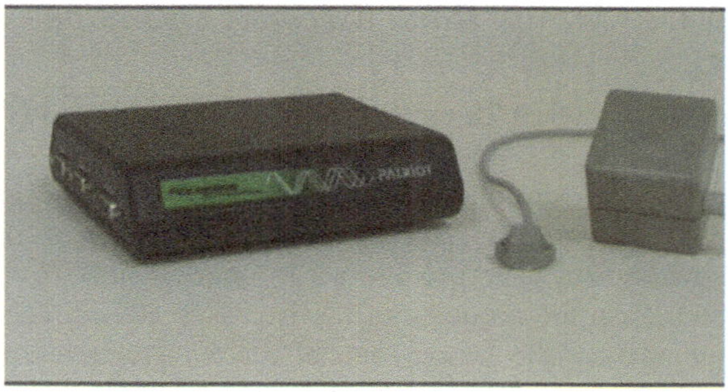

Fig. 6.9 Patriot location tracker

Before using the position tracker, you first need to place the sensor on the origin of the coordinate system of the coordinate source, and reposition the sensor to ensure that the coordinates of the two are consistent. Here we use the position tracker Patriot. Patriot is mainly composed of a position sensor, a fixed coordinate source and a signal converter. Before use, you need to place the sensor on the origin of the coordinate system and reposition the sensor to ensure that the coordinates of the two are consistent. The Patriot position tracker is as shown in Fig. 6.9.

Wear the data glove on the right hand, and then insert the sensor part of the position tracker into the palm part of the data glove, which can transmit the position of the palm to the OSG virtual reality environment through the position tracker. The structure diagram and the bending of the fingers, so that the relationship between the position tracker and the data glove can be established, and the purpose of the operator's hand driving the virtual hand in the virtual reality environment is achieved. Figure 6.10 shows the operator wearing the data glove and position tracking the state of the operator's hand, the movement of the operator's hand and the bending of the fingers during the test can be truly reflected on the screen.

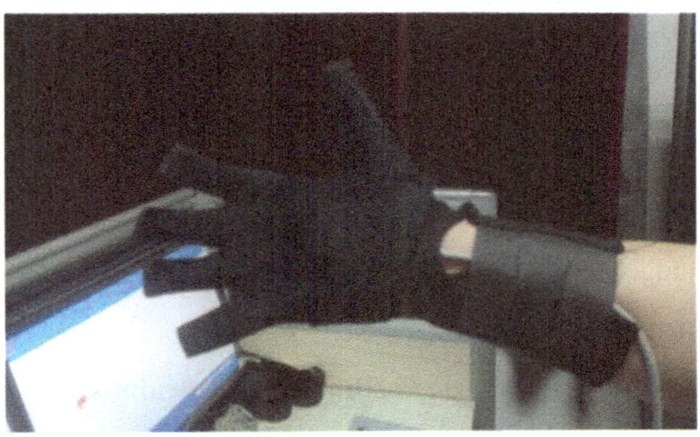

Fig. 6.10 Wearing gloves

6.3.4 *Virtual Hand Assembly Operation*

6.3.4.1 Research on the Rules of Virtual Hand Grasping

The virtual hand mainly performs virtual assembly operations by grasping, operating, and releasing. Therefore, the rules for grasping parts of the virtual hand in the virtual environment have become a key issue, mainly from when to successfully grasp and grasp the accuracy of the two aspects is discussed. This chapter divides the judgment of the grasping model into two processes: the recognition of gestures and the realization of grasping rules. That is, firstly, the gesture recognition is used to determine whether the operator has the intention of grasping the object, and if so, the manual grasping rule is used to determine whether the object is successfully grasped.

When a human hand model is driven in a virtual scene, it is necessary to judge when the object to be assembled is successfully captured. The accuracy and fidelity of the captured model is also the key to ensuring a strong sense of reality in the virtual scene.

The SDK library that comes with Data Glove can recognize and judge some basic gestures, which can be obtained through fdGetGesture(). This method is to obtain 16 gestures by comprehensively judging the degree of bending of 4 fingers (except the thumb). In actual operation, a threshold is set for the bending of each finger joint. If the collected finger bending value is greater than the threshold, then It is considered that this finger is curved. Since a total of 4 fingers are considered and each finger has 2 possible situations, there are a total of 24 gesture situations. This chapter uses fdGetGesture() == 0, the first four-finger gripping style, when judging gestures.

If the operator grips with four fingers, the system will judge that the operator has the intention of grasping the object, and the grasping plan needs to be studied to determine whether the object is successfully grasped, and the ray intersection method is used to make the judgment. Create OSGUtil::LineSegmentIntersector at the positions of the five fingers. When the ray of the finger intersects the surrounding sphere of the component, it is the contact with the object. According to the actual grasping law of the human hand, when the three fingers including the thumb are The finger touches the object and all bends below 2/3 of the entire palm plane, that is, grabbing the object.

The release of the virtual hand is contrary to the grasping rules. When three or more rays are far away from the surrounding ball and do not intersect with it, it is judged to release the object. The grasping release rule diagram is shown in Fig. 6.11.

6.3.4.2 Automatic Positioning Constraints

When the virtual hand releases the object, the position of the object is also a key issue. When the virtual hand is released, first record the position coordinates at this time, and then make a judgment. If it and the correct assembly position are greater than a certain value, the object will stay at the coordinates at this time. If it and the

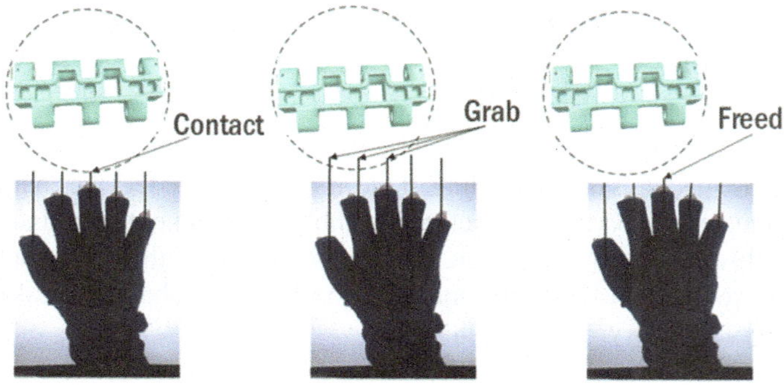

Fig. 6.11 Grab release rule diagram

correct assembly position are less than a certain value, the object will be automatically assembled to the correct position to complete the assembly.

6.3.5 *Implementation of Assembly Interaction Based on Virtual Hands*

Based on the content discussed above, this chapter implements another assembly interaction method of the system—the assembly interaction based on data gloves and position tracker, as shown in Fig. 6.12. The figure shows the virtual hand assembly interaction for traction outside the shearer.

Fig. 6.12 The virtual hand assembly interaction of the external traction of the shearer

6.4 Force Feedback Human–Computer Interaction Subsystem

In the virtual assembly process, if the sense of force and touch can be added, it will further increase people's understanding of the assembly parts, feel the characteristics and constraints of the parts in the actual assembly environment, and find problems in the assembly process in time, such as there are defects in the size structure. This chapter is developed based on the force feedback device phantom desktop in the virtual reality laboratory.

6.4.1 Introduction to Phantom Desktop Equipment

Figure 6.13 shows the structure diagram of the force/tactile feedback application system. The operator, the force feedback device and the computer-generated virtual environment constitute a two-way closed-loop system. The working principle of the system is as follows: First, the operator completes the input of 6 degrees of freedom by operating the force feedback device, and the operator only directly operates the handle at the end of the device. The position of the force/tactile interface point HIP is shown on the left of Fig. 6.14. Point P in the figure. Then the space matching module completes the pose conversion, and maps the pose input by the user to the pose of the proxy point in the virtual environment (the Q point in the right of Fig. 6.14), and the pose conversion is completed through the space matching algorithm [5].

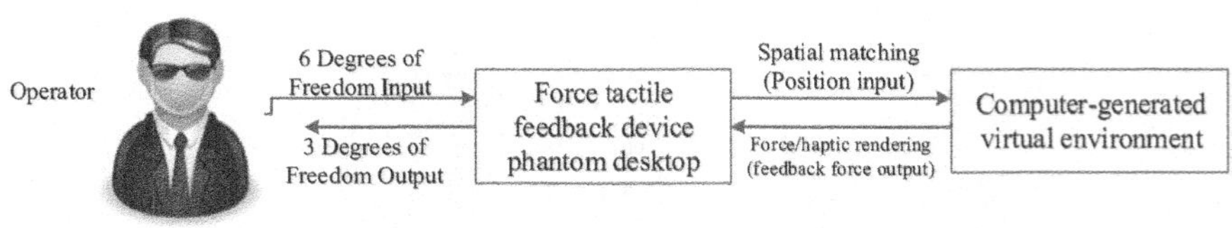

Fig. 6.13 Structure diagram of force/tactile feedback application system

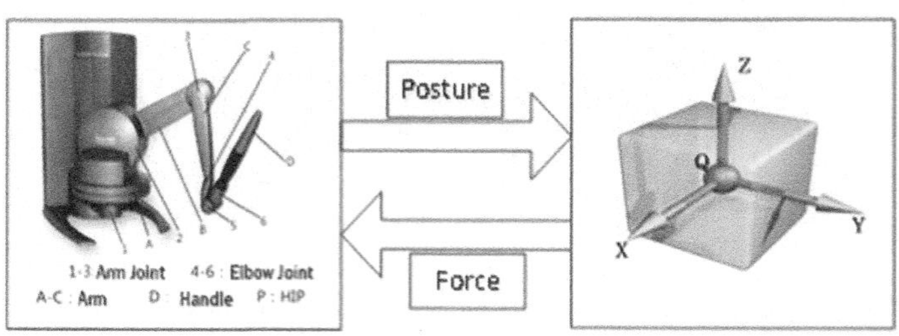

Fig. 6.14 Phantom desktop structure diagram and interaction schematic diagram

6.4.2 *Subsystem Frame Design*

In the process of the agent point (HIP point) interacting with the virtual environment, there are mainly the processes of picking up, changing the pose and releasing. To complete the interactive process, we must first pick up the target object. Press the button on the force feedback device, assuming that the position of the proxy point collides with various parts of the virtual environment, then the virtual part that collided will be picked up, thus completing the picking work of the virtual model. This virtual model becomes the object currently being manipulated. After completing the manipulation of the current part, release is next. The operator releases the button on the force feedback device, then the currently operated virtual part can be released to the current position. Through the operations of the above three processes of picking up, changing the pose and releasing, the interaction between the agent point and the virtual environment can be completed [6].

6.4.3 *Model Import Technology*

Since the programming interface of the force feedback device phantom desktop is the openGL environment, this chapter needs to import the established model into the openGL environment. There are two ways to import, one is to directly use the openGL language to model the simple model, and the second One is to use the obj format as an intermediate format for importing. The two importing methods are shown in Fig. 6.15.

- A relatively simple model can be directly modeled using the openGL modeling language, for example: glutSoildSphere(0.5,32,32) //Create a physical sphere
- The existing coal machine equipment models are UG or Pro/E models. Import these two models into 3DMAX using the intermediate format STL, and then export the obj format file in the 3DMAX software. By writing a program, you can convert the obj Import the file into the openGL environment.

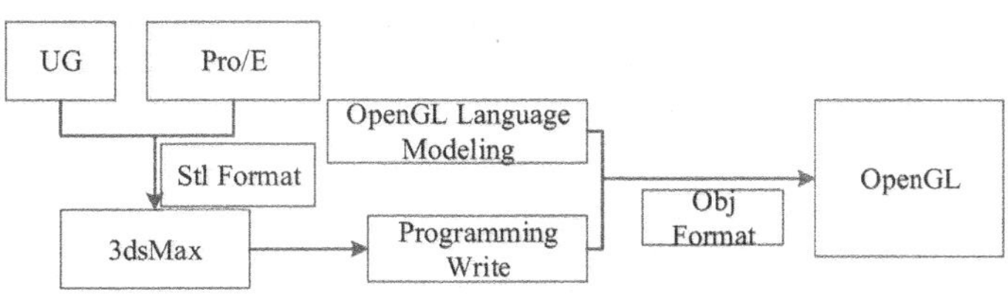

Fig. 6.15 Method of importing 3D model into virtual environment

6.4.4 *Force Rendering of Objects*

The model has a series of properties. The structure structDraggableObject of the object to be retrieved is established, as shown below:

```
structDraggableObject//The structure of the object to be grabbed
{HLuintshapeId;//Tactile state
GLuintdisplayList;//Object list
hduMatrix transform;//transformation matrix};
```

When an object is touched, to have a sense of force, it is necessary to perform force/tactile rendering on the face of the object. You only need to call the following function in the main thread:

```
hlBeginShape(HL_SHAPE_FEEDBACK_BUFFER, obj.shapeID);//Start force effect
glCallList(obj.displayList);//Read the list of objects
hlEndShape();//The force sense drawing is completed
```

Objects have many forms of force effect. Here, the contact model is used as contact. When the surface is touched, force sense is activated. The following function is used to achieve this function:

```
hlTouchModel(HL_CONTACT);//Set the contact model to contact
hlTouchableFace(HL_FRONT);//Set the surface to be touched
```

6.4.5 *Principle of Force Feedback Control Model*

A virtual interaction process usually includes at least three stages: picking up, changing the pose, and releasing. If the current proxy point collides with a virtual model and the operator presses the button on the phantom arm, the virtual model will be picked up and become the currently operated part, and the coordinate system will appear and follow the grasped object Move together; when the operator releases the button on the phantom arm, the currently operated part will be released to the current position.

6.4.6 *Automatic Positioning Constraints*

When the button is pressed and the agent point touches the object, the object is grabbed and moves along with the agent point. When the button is released, the position of the object will be released to the position of the current agent point, and the program will compare the current position with the correct assembly position of the object is judged. If it is less than the constraint, it will be automatically assembled

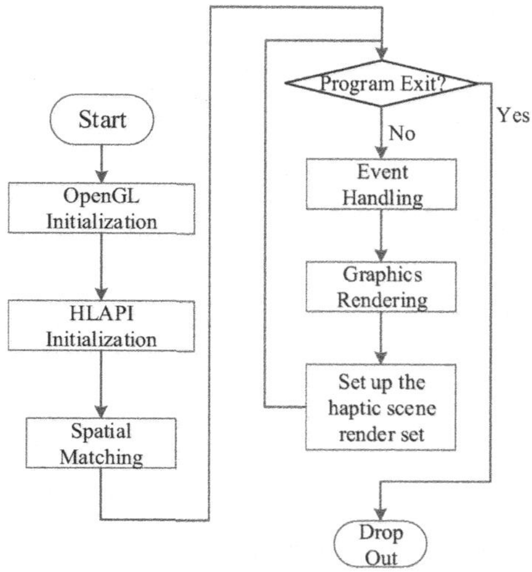

Fig. 6.16 HLAPI main thread framework

to the correct position. If it is greater than the constraint, the object is still at the release position of the proxy point.

6.4.7 Tactile and Visual Rendering Modes

This module is developed using HLAPI combined with HDAPI, including two threads of force and vision. Coordination between the two must be done, otherwise it will cause a mismatch between force and vision. The program framework is shown in Fig. 6.16.

References

1. Fenfang, Z., Liang, B., Liu, Z., et al.: Human-computer interaction environment design based on data gloves. Chin. J. Image Graph. **5**(2), 153–156 (2000)
2. Lin, L., Wu, P., Xiong, W.: Research on the key technology of virtual supermarket cognitive rehabilitation training system development. J. Graph. **35**(01), 105–109 (2014)
3. Zhou, X., Zhao, Z., Lou, J.: Virtual gesture interaction system based on data gloves. Instrum. Technol. Sens. **10**, 65–70 (2007)
4. Iwasako, K., Soga, M., Taki, H.: Development of finger motion skill learning support system based on data gloves. Procedia Comput. Sci. **35**(6), 1307–1314 (2014)
5. Cui, Y., Wang, T.: Nail hole assembly based on force feedback in virtual environment. J. Changchun Univ. Sci. Technol. (Nat. Sci. Ed.) **02**(34), 100–104 (2011)
6. Yuan, X.B., Yang, S.X.: Virtual assembly with biologically inspired. IEEE Trans. Syst. Man Cybern. Part C Appl. Rev. **33**(2), 159–167 (2003)
7. Lu, L., Liu, L., Xiong, W.: Research on virtual assembly based on single-point force feedback device. J. Graph. **35**(02), 280–284 (2014)

Chapter 7
Virtual Assembly Network Method

7.1 Introduction

Sharing virtual reality resource library resources such as virtual disassembly and scene simulation through the network to provide resource sharing and technical support for coal mining machinery equipment companies, especially small and medium-sized enterprises, is an important driving force for the establishment of coal mining equipment virtual reality assembly network technology and systems. Aiming at the field of coal machinery equipment, with the goal of improving design and use innovation capabilities, with the establishment of a sharing mechanism as the core, making full use of modern network information technology, integrating and optimizing scientific and technological resources for the virtual design of coal mining machinery, and building public welfare, basic and independent intellectual property rights The virtual reality assembly network technology and system of coal mining machinery equipment mainly provides and realizes convenient virtual disassembly and assembly of coal machinery equipment and typical scene simulation services.

7.2 Subsystem Frame Design

"Coal Machinery Equipment Virtual Reality Assembly Networking Technology and System", based on the browser/server (B/S) model, implements collaborative functions in a collaborative environment with specific application modules, and completes effective data with model management, the system architecture is shown in Fig. 7.1:

- Client: Through the use of WWW technology, the application of HTML, ASP and other Web pages, VBScript and other programming languages and ActiveX controls are combined to provide users with a graphical user interface. Through the interface, the client user can complete the operation and display of virtual reality resources to achieve Interaction with the web server.

X. Wang et al., *Virtual Reality Technology in Mining Machinery*,
https://doi.org/10.1007/978-981-16-4408-5_7

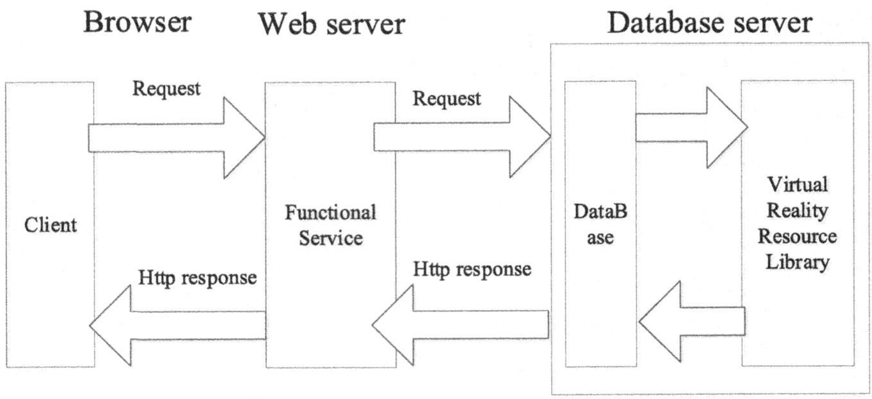

Fig. 7.1 Subsystem architecture

- Functional services: Mainly the functional modules on the server side to realize application logic such as access and retrieval of virtual reality resources. It is the core of the system, including virtual disassembly and assembly modules, scene simulation modules, document modules, etc., to achieve analysis and Data management, etc.
- Database server: Provides data resources such as models, instances, and resources in the process of virtual reality services.

7.2.1 Subsystem Hardware Design

This subsystem aims to realize remote sharing of existing virtual disassembly and scene simulation resources through the network. It adopts the B/S mode. Users only need to open the browser on their laptop or desktop computer and enter the URL to enter the system. After registering, users will get installation instructions and simply configure their own computer terminals to enjoy the functions provided by this system. To this end, the required server and client configurations are as follows:

- Server: One configured with IIS, the system is Win7 or WinXP, 32-bit or 64-bit high-performance computer;
- Client: The system is Win7 or WinXP, 32-bit or 64-bit computer, equipped with IE or several other browsers.

7.2.2 Subsystem Software Design

During the establishment of the system platform, four main problems were solved, namely the establishment of the virtual reality database, the design of the back-end

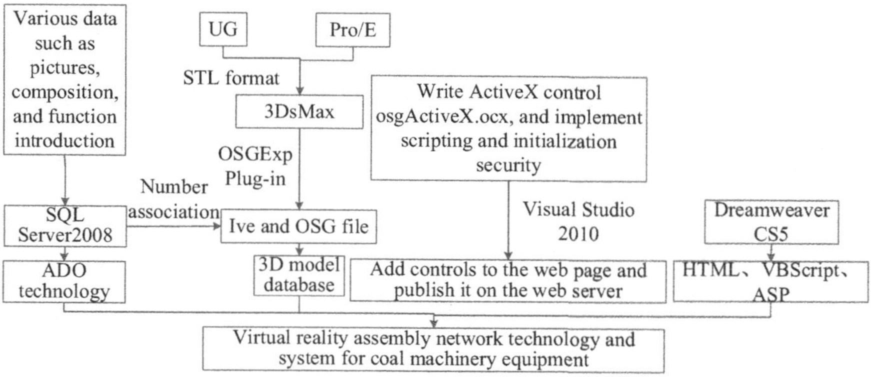

Fig. 7.2 Software system design and key technologies

database, the design of the website interface and the display of the virtual reality files on the website, the establishment of the software system and the software used in the system. The key technology is shown in Fig. 7.2.

Use UG, 3DMAX, OSG (Open Scene Graph) to build a virtual reality resource library, Dreamweaver CS5, ASP and VBScript to build a basic network site, use ActiveX controls to display virtual reality resources in web pages, and build on the basis of ADO technology and SQL server2008 The back-end database is used to develop practically valuable coal machine equipment virtual reality assembly networked technology and systems.

7.2.3 *Subsystem Structure Design*

This system is centered on the virtual disassembly and assembly demonstration of coal machinery equipment structure and typical scene simulation, and integrates other resources such as coal machinery equipment literature, lecture halls, etc., to provide users with services. Users can register by themselves and select related services. The structure diagram is shown in Fig. 7.3.

7.3 Virtual Reality Resource Library

The virtual reality resource library mainly includes coal machine equipment virtual disassembly and assembly resources and scene simulation resources. This part mainly relies on the virtual reality resource library described in the third chapter of this book, so it won't be repeated here.

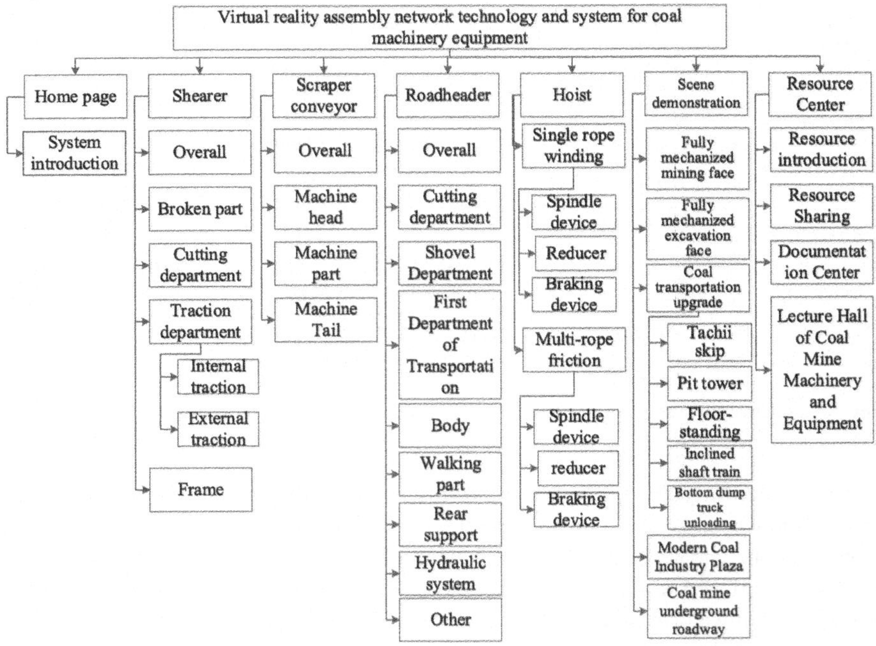

Fig. 7.3 System structure diagram

7.4 ActiveX Control Technology

Using ActiveX control technology, the files of the virtual reality resource library can be realized in the web page. The purpose is to write an MFC ActiveX control, in which to realize the display and data interaction of the OSG window, and use the method of registering OCX to embed the OSG window into the browser to realize the 3D model display function on the web page.

7.4.1 Writing OSG-ActiveX Controls

The main ideas and steps for writing OSG-ActiveX controls are as follows:

- Get the handle of the MFC window, and create a new graphics device context GC (Graphics Context) accordingly;
- Create a camera and specify the GC, Viewport and perspective matrix used by it;
- Set the camera as the main camera of the Viewer class (Viewer) to fit the OSG scene into the designated MFC window;
- OSG's rendering loop should not be placed in the OnDraw or OnPaint function of MFC, because these two functions can only receive messages when the window

needs to be redrawn, so OSGViewer::Viewer::frame cannot be executed in it. Render the scene. At this point, a new thread can be established, the rendering loop can be implemented in the new thread, and the thread can be terminated in time when the program exits.

7.4.2 *Server-Side Control Release*

Implementation of HTML code on the browser side. To observe the OSG scene in the browser, you first need to load the compiled OCX control. The general method is: enter the Windows command line mode, and enter regsvr32 OSGActiveX.ocx, the system will prompt that the control has been loaded. The command to uninstall the control is: regsvr32/u OSGActiveX.ocx. Use the Windows built-in VC decompiler tool (depends.exe) to open the ocx file, and copy the DLL file required by this control to the system folder (system32), so as not to find the necessary dynamic link library when the control is running. Then use a relatively simple method in the HTML code to register the OCX control immediately, so as not to bring unnecessary trouble to the debugging of the program. The main code is as follows:

```
< OBJECT
classid = "clsid:9AB36F74-9505-4B3E-A9D6-6294F67C804D"
id = OSGOcx codebase = OSGActiveX.ocx width = 1000
height = 680 >
< /OBJECT > < BR >
```

Note that the classid here is obtained from the IMPLEMENT_OLECREATE_EX line of the source file. This completes the release of the control on the server side.

7.4.3 *Client Environment Configuration*

The dotted line of real resource files is relatively large. If you put it on the server side, on the one hand, the network transmission speed is slow, and on the other hand, it causes huge pressure on the server side, which may cause a system crash. So this system uses FTP to transfer virtual reality resources Download it, name it according to the requirements of the background database, put it in the specified folder as required, and add this file directory to the computer environment variable.

For example, add the path C:\ProgramFiles\OpenSceneGraph\data to the environment variable OSG_FILE_PATH, the code is as follows:

OSG_FILE_PATH = "C:\ProgramFiles\OpenSceneGraph\data".

7.5 Basic Interface Design

Use ADO technology to connect to the database, create conn.asp in the main directory conn folder, the code is as follows:

```
Dim Conn
Set Conn = Server.CreateObject("ADODB.Connection")
Conn.Open
"Driver = {Sql Server};
Server = (local);UID = sa;PWD = ******;Database = xiandaisheji"
```

This completes the connection with the database.

Figure 7.4 is the virtual disassembly and assembly module resource list of the scraper conveyor. It displays all the resources with the number 1 and the value 2 in the xunizhuangpei table in the database in the form of a list, arranged according to ID from small to large, each page has 6 columns. Bar, which is convenient for users to view resources according to their own needs. The database query statement established is as follows.

```
Sql = Select * from meijizhuangbei where bianhao1 = 2 order by ID Asc.
```

There is also a search function on the left side. Users can choose to enter a code or a keyword to obtain the corresponding qualified resources, and realize the visual data query, which is mainly realized by the following code:

```
< option value = "bianhao1" < %If(InStr(Request("condition"), "bianhao1") > 0)Then% > selected < %End If% > > Number One < /option >
< option value = "name" < %If(InStr(Request("condition"), "name") > 0)Then% > selected < %End If% > > Name < /option >
...
< input name = "Key" type = "text" style = "width:100px;border:1px solid" value = " < % = Server.HTMLEncode(Request("Key"))% > " >
```

7.6 Back-End Database Design

Using SQL SERVER2008 as the back-end database software of this system, a total of five tables are designed, virtual disassembly module table (xunizhuangpei), scene simulation module table (changjing), administrator and user login table (User), coal machine equipment information table (meijizhuangbei), Document Information Table (wenxian), Table 7.1 is a virtual disassembly module table.

The meaning of the numbers 1–4 is the corresponding level and level code. Number 1 is the highest level. When the value is 1, it represents a shearer, 2 represents a scraper, 3 represents a roadheader, and 4 represents a hoist. When the value of number 1 is 1, the value of number 2 is 1 for the complete machine, 2 is the crushing part, 3 is the cutting part, 4 is the walking part, and 5 is the machine frame.

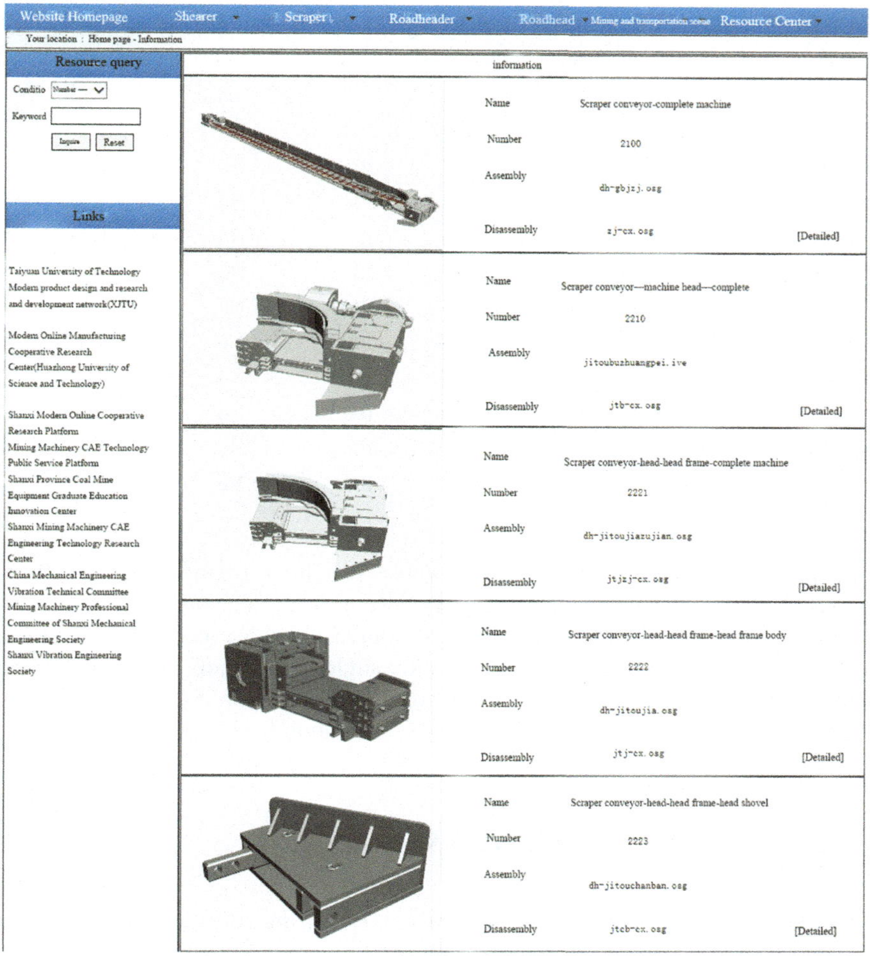

Fig. 7.4 Resource list of scraper conveyor module

7.7 Public Service Edition

Due to the large size of virtual reality files, directly placing them on the server side will cause network congestion and other problems. Downloading to the client side for application also involves confidentiality issues. This is a pair of contradictions that urgently need to be further studied and resolved.

Due to the above problems, it is suggested that this Web-based system can be divided into two categories. One: For internal use in the enterprise, the enterprise has fully obtained the authority, and can use ActiveX controls to operate the

Table 7.1 Virtual disassembly module table design

Name	Field Name	Type of Data	Data Length	Primary Key	Non Empty
Serial Number	ID	Int	–	Yes	Yes
Name	Name	Varchar	200	No	Yes
Number 1	Bianhao1	Varchar	20	No	Yes
Number 2	Bianhao2	Varchar	20	No	Yes
Number 3	Bianhao3	Varchar	20	No	Yes
Number 4	Bianhao4	Varchar	20	No	Yes
Total Number	Bianhao	Varchar	100	No	Yes
Composition	Zucheng	Varchar	2000	No	No
Features	Gongneng	Varchar	2000	No	No
Picture Name	Img_name	Varchar	100	No	No
Assembly File Name	zhuangpei	Varchar	100	No	No
Disassemble File Name	chaixie	Varchar	100	No	No

corresponding virtual reality resources, mouse switching, rotation, etc.;: For non-cooperative enterprises or individual users, virtual reality resources involve confidentiality. Therefore, for this virtual reality resource, use video recording software to make the corresponding video from a specific angle. Specifically, the virtual disassembly and assembly part is used to record the assembly process from a specific angle. And scene simulation is to make scene animation from a specific angle.

7.7.1 Choose Video Production Software

First choose a good video recording software. Either the software on the existing network does not work well, or there is no need to pay. After using about ten software, it is found that the "Ruidongtiandi" super video recording software meets the requirements and the effect is good.

7.7.2 Select the Format to Be Played

There are mainly flash, mpg, AVI, which can also be achieved by using the video conversion software of "Ruidongtiandi". After comparison, the same video has different file sizes and different characteristics:

- The MPG file is larger and clearer;
- The AVI is relatively small and the clarity is better;
- The flash is relatively small and the definition is average.

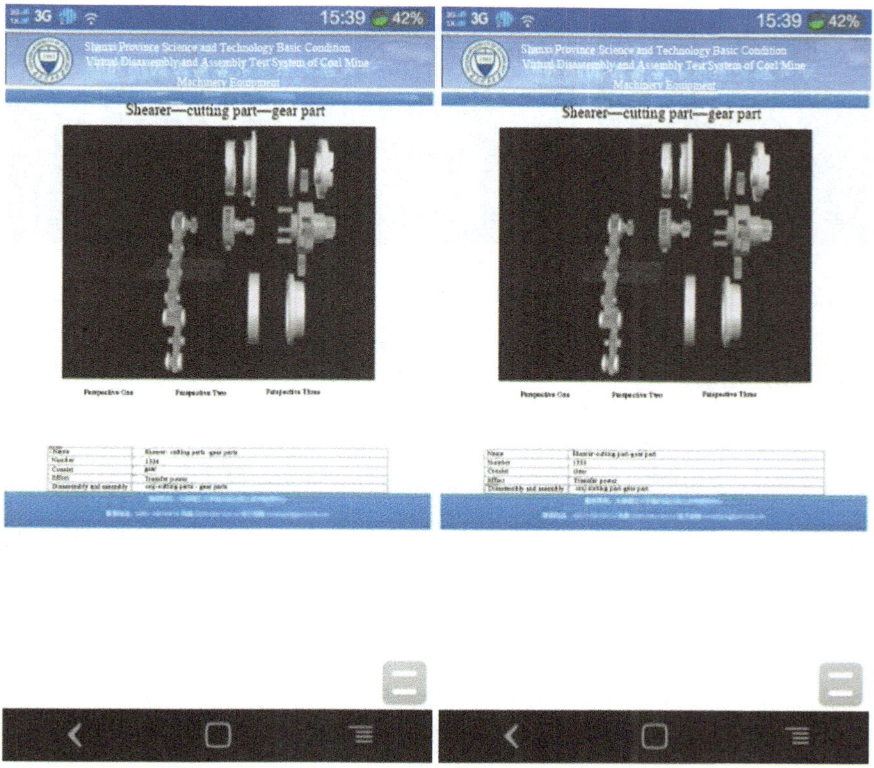

Fig. 7.5 Huawei mobile phone test renderings

7.7.3 Network Playback Code and Effect Test

We searched for flash, mpg, AVI and other playback codes on the Internet, and tested various video effects. It can be seen that the effect of AVI is better, mpg is the best, and flash is worse. However, flash can be played on some mobile phones (Apple mobile phones cannot be used, Samsung and others may need to download plug-ins, and Huawei and other mobile phones can be played directly). Figure 7.5 shows the test results on Huawei mobile phones.

7.7.4 Multi-view Playback

Recording each scene from three different angles can make the user feel the simulation scene of each assembly and equipment clearest to the greatest extent. Figure 7.6 shows the playback effect diagram of three perspectives:

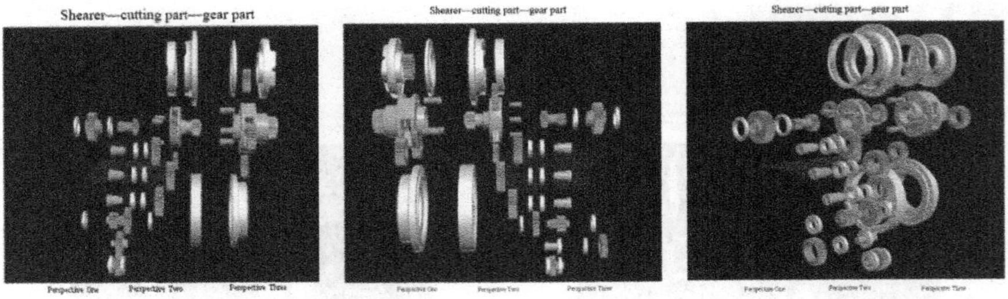

Fig. 7.6 Play effect diagram from three perspectives

7.8 Summary

Proposes the operation mode of the virtual assembly and simulation system based on Web. Aiming at the characteristics of the two different user groups of the enterprise and the public, two systems for the internal and public services of the enterprise are built respectively, which solves the difficulty of accessing virtual reality resources in different places.

Chapter 8
Method and Technology of Virtual Single Machine Simulation

8.1 Introduction

In order to study the monitoring and planning methods of "three-machine" working condition in fully mechanized coal face under VR environment, it is necessary to study the monitoring and virtual simulation method of single machine working condition in "three-machine". In the working condition monitoring part, the actual postures of shearer, scraper conveyor and hydraulic support under the condition of underground condition needs to be monitored by sensors arranged at a specific position on each machine. In the part of the virtual simulation method, the research of virtual machine simulation method is carried out by using Unity3D virtual reality simulation engine, which is in accordance with the actual equipment behaviors.

8.2 Establishment of Physical Information Sensing System

This book takes the "Three-machine" in fully mechanized coal face t of the complete set of test system of coal mine fully mechanized mining equipment of Shanxi Key Laboratory of Fully Mechanized Coal Mining Equipment as the research objects. The models of "three-machine" and other information are shown in Table 8.1.

8.2.1 Sensor Layout of Shearer

In order to satisfy the needs of panoramic shearer attitude monitoring, the sensor layout scheme of shearer is shown in Fig. 8.1. The functions of each sensor are as follows:

X. Wang et al., *Virtual Reality Technology in Mining Machinery*,
https://doi.org/10.1007/978-981-16-4408-5_8

Table 8.1 Overview of complete set of test system equipment for fully mechanized coal mining

Number	Equipment name	Specification and model	Number/set
1	Hydraulic support	ZZ4000/18/38	20
2	Shearer	MGTY250/600	1
3	Scraper conveyor	SGZ764/630	1

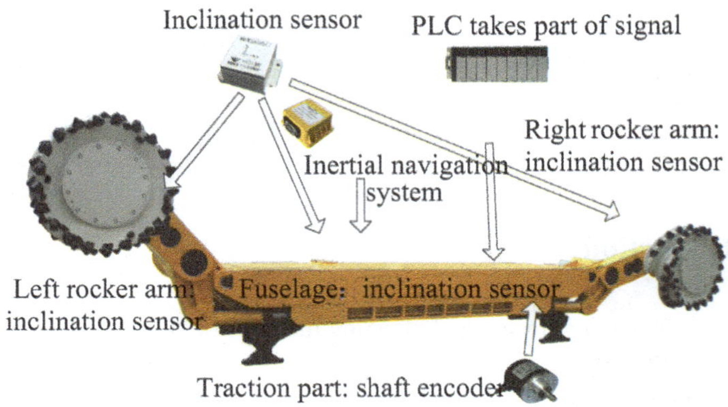

Fig. 8.1 Sensor layout of shearer

- Shaft encoder of traction part: used for rough positioning of shearer;
- Fuselage inclination sensor: it is used to monitor the inclination angle of the shearer fuselage. The fuselage inclination sensor monitors the pitch angle and roll angle of the shearer. When the terrain conditions change, the cutting height can be compensated in the process of memory cutting to achieve stable and accurate memory cutting;
- Rocker angle sensor: it is used to monitor the left and right rocker angles of shearer, perform real-time calculation with the shearer body angle, and solve the absolute rotation angle of the hinge point between the two rocker arms and body;
- Strapdown inertial navigation: the inertial measurement device is directly fixed on the center of the shearer body for navigation solution. The strapdown inertial navigation device is composed of nine measuring units including three-axis gyroscope, three-axis accelerometer and three-axis magnetometer, which can accurately locate the shearer. It can output the roll angle, pitch angle and yaw angle of the shearer in real time, and fuse with the sensor information of the fuselage inclination angle in real time, so as to improve the measurement accuracy.

All the signals return to the central control center through the RJ45 signal port of the airborne PLC and the mine intrinsically safe wireless base station.

8.2.2 Sensor Arrangement of Hydraulic Support

The sensor layout of hydraulic support is shown in Fig. 8.2. In order to meet the requirements of attitude monitoring of panoramic support shield type hydraulic support, four dual axis tilt sensors should be arranged at the shield beam, front link, base and top beam respectively, and proximity sensors should be arranged at the primary and secondary protection and telescopic beam respectively. All signals can be connected to the network I/O acquisition module and returned to the central control center through the high-speed network communication platform. The data of the monitoring programmer are as follows:

- Top beam inclination angle: the top beam elevation angle is calculated in real time in combination with the base inclination angle. Generally, the elevation angle shall not exceed ±7°;
- Support height of support: the support height of support is calculated by the inclination sensors arranged at the top beam, shield beam, front connecting rod and base;
- Attitude of hydraulic support: through the tilt sensors arranged at the top beam, shield beam, front link and base, the telescopic length of four columns can be calculated respectively through attitude calculation, and then the attitude of the whole support can be obtained;

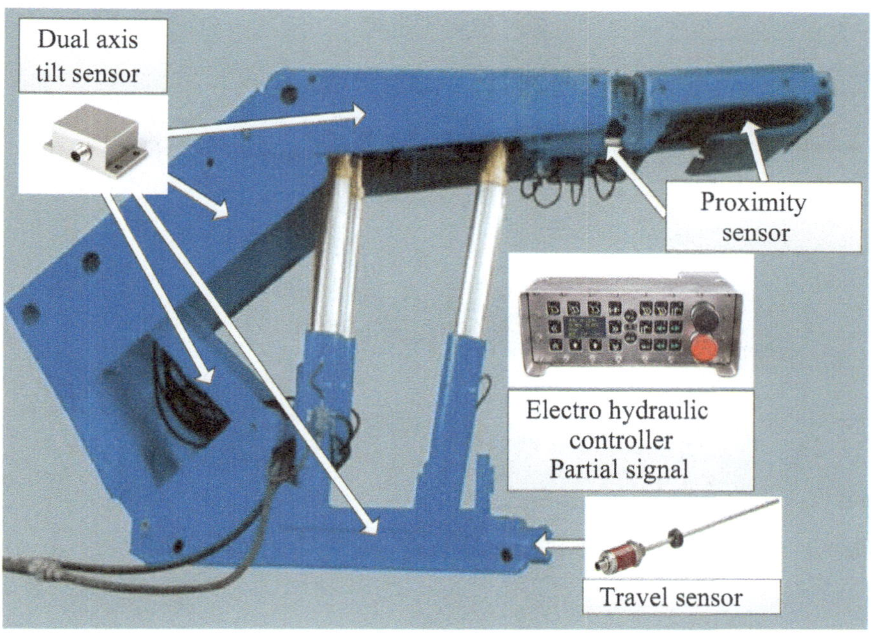

Fig. 8.2 Layout of sensors on single hydraulic support

- Whether the primary and secondary mutual aid and telescopic beam are in place: close to the sensor, the signal changes after the telescopic beam is in place;
- Displacement and state of bracket pushing oil cylinder: displacement sensor of pushing oil cylinder is used to judge bracket pushing state;
- Bracket arrangement status: there are many kinds of bracket arrangement status, such as whether two adjacent brackets bite, whether the brackets keep straight, etc. The arrangement state of the support can be calculated synthetically by solving the attitude of the single support.

8.2.3 Sensor Arrangement of Scraper Conveyor

The sketch of scraper conveyor terrain monitoring is shown in Fig. 8.3. In order to monitor the shape of the working face floor, it is necessary to arrange double axis inclination sensor on each scraper conveyor chute. The specific installation position is the lower side of cable trough of each chute, which can realize the real-time measurement of the double axis inclination of each chute, and complete the analysis of the terrain state of the comprehensive mining face. The attitude of each chute was collected, and the 3D terrain of the working face was restored by the analysis and summary of the calculation method. In the process of coal mining, the equipment attitude of the working face is adjusted timely by controlling the undercover, mining height and hydraulic support bed.

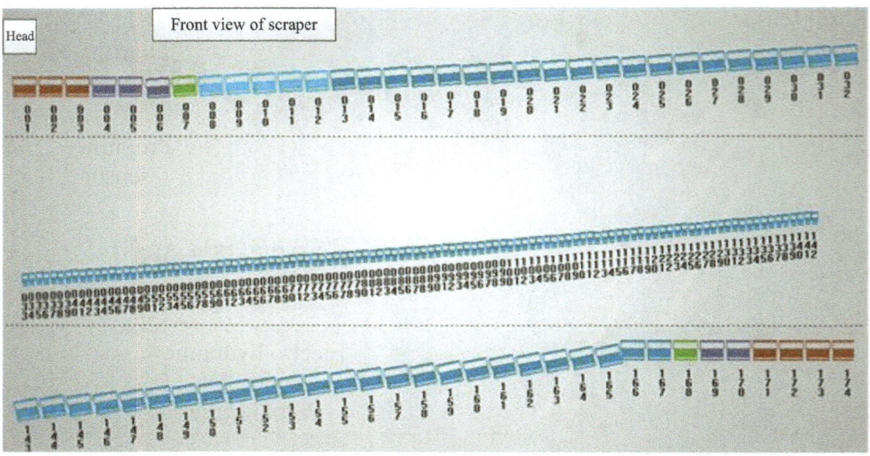

Fig. 8.3 Attitude monitoring scheme of scraper conveyor

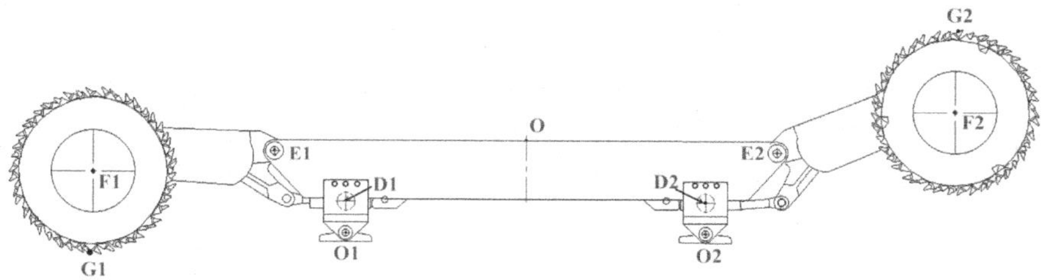

Fig. 8.4 Key points analysis of shearer. O1—key point of left support slipper; O2—key point of right support slipper; D1—key point of left guide wheel; D2—key point of right guide wheel; E1—hinge point of left rocker arm; F1—rotation point of left roller; E2—hinge point of right rocker arm; F2—rotation point of right rocker arm; G1—calculation point of right roller; G2—calculation point of right roller

8.3 Single Machine Attitude Monitoring Method of Fully Mechanized Mining Equipment

8.3.1 Single Machine Attitude Monitoring Method of Fully Mechanized Mining Equipment

These key points can be calculated in real time (Fig. 8.4) to describe the attitude of the shearer.

When there is a longitudinal inclination angle, the key point coordinates can be easily calculated by converting according to the longitudinal angle. The attitude monitoring of shearer mainly calculates the accurate position of shearer through the positioning of strapdown inertial navigation system, so as to determine the coordinates of o point. Then, through the angle information of the fuselage angle sensor and the angle information of sins, the accurate roll angle, pitch angle and yaw angle of the shearer can be calculated in real time. Then through the left and right rocker angle sensor, the absolute rotation angle of the hinge point between the two rocker arms and the fuselage is solved, so as to fully express the attitude of the shearer.

8.3.2 Attitude Monitoring Method of Scraper Conveyor

As shown in Fig. 8.5, it is known that the length of each section of the middle trough is LZBC, and the lateral inclination of each middle trough is α_n, The longitudinal inclination is β_n. It can be seen that the segment function of each central trough in XY plane (Formula 8.1):

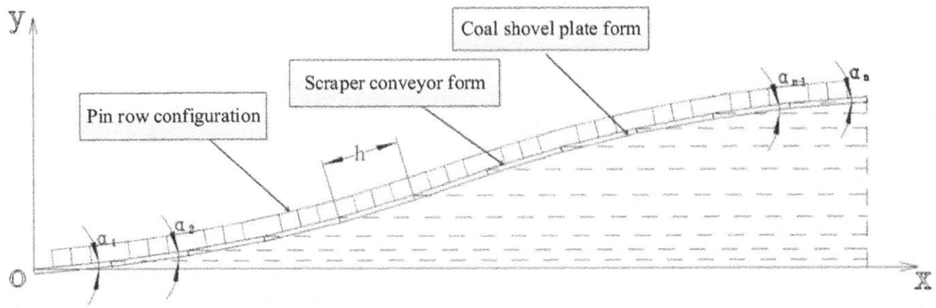

Fig. 8.5 Scraper conveyor form

$$\begin{cases} f_1(x) = x\tan\alpha_1, & 0 \le x \le x_1 \\ f_2(x) = f_1(x_1) + (x - x_1)\tan\alpha_2, & x_1 < x \le x_2 \\ \quad \ldots \\ f_{n-1}(x) = f_{n-2}(x_{n-2}) + (x - x_{n-2})\tan\alpha_{n-1}, & x_{n-2} < x \le x_{n-1} \\ f_n(x) = f_{n-1}(x_{n-1}) + (x - x_{n-1})\tan\alpha_n, & x_{n-1} < x \le x_n \end{cases} \tag{8.1}$$

where x_i is the boundary point of the i-th middle slot on the X-axis.

The position of shearer on scraper conveyor, the key point O_1 of shearer is located at p on the k section of scraper conveyor.

$$\frac{s}{h} = k \cdots p \tag{8.2}$$

where s is the stroke of the shearer, h is the length of the middle slot of a single section, k is the quotient, representing the number of the middle slot of the shearer, and p is the remainder, representing the p-th position of the shearer in the k-th middle slot.

The coordinates (x_k, y_k) of the k-th hinge point of the scraper conveyor and the coordinate offset (x_k, y_k) of p on the k-th section of the scraper conveyor relative to the point (x_p, y_p) are obtained. Therefore, when the stroke of the shearer is s, the coordinate of the characteristic point of the shearer is:

$$\begin{cases} x_s = x_k + x_p = L_{ZBC}\sum_{i=1}^{k}\cos\alpha_i + L_{ZBC}\cdot p\cdot\cos\alpha_{k+1} \\ y_s = y_k + y_p = L_{ZBC}\sum_{i=1}^{k}\sin\alpha_i + L_{ZBC}\cdot p\cdot\sin\alpha_{k+1} \end{cases} \tag{8.3}$$

8.3.3 Attitude Monitoring Method of Hydraulic Support

8.3.3.1 Attitude Analysis

The attitude analysis of the hydraulic support mainly includes the analysis of the four-bar linkage, the cooperative analysis of the four-bar linkage and the top beam, and the cooperative analysis of the four-bar linkage and the top beam and the front and rear columns.

The analysis shows that: given the front link angle or the rear link angle, the attitude of the four-bar mechanism is determined, including the attitude of the shield beam, but the movement of the top beam is affected by the movement of the shield beam. Combined with the working condition of the roof, the pitching action is made. Therefore, the top beam also has an independent degree of freedom, and the attitude parameters of the whole hydraulic support can be obtained by using the variable correlation of the top beam angle. Establishing the analytical model of hydraulic support as shown in Fig. 8.6.

In this section, the attitude is analyzed based on the known inclination angles of front link and top beam. The angle variable information established is shown in Table 8.2.

Given the structural parameters of L_1, L_2, L_3, L_4, L_5, θ and φ, for ZZ4000/18/38 hydraulic support, $L_1 = 379.6$ mm, $L_2 = 1375.4$ mm, $L_3 = 1400$ mm, $L_4 = 686.4$ mm, $L_5 = 190.5$ mm, β and γ are set.

It can be seen from the analysis of the relationship in the figure that:

$$\begin{cases} L_2 \sin\beta + L_4 \sin\phi = L_1 \sin\gamma + L_3 \sin\theta \\ L_2 \cos\beta + L_1 \cos\gamma = L_4 \cos\phi + L_3 \cos\theta \end{cases} \quad (8.4)$$

Solutions:

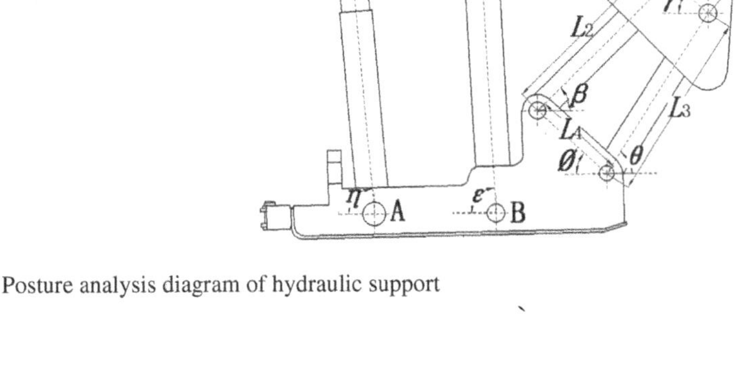

Fig. 8.6 Posture analysis diagram of hydraulic support

Table 8.2 Angle variable of ZZ4000/18/38 hydraulic support

Letter	The meaning of the letter
θ	Rear linkage angle
δ	Top beam inclination
φ	Inclination angle between connecting line of front connecting rod pin and rear connecting rod and base
β	Front link inclination
γ	Angle of shield beam
η	Angle between front column and base
ε	Angle between rear column and base

$$\gamma = \arcsin\left(\frac{c}{\sqrt{a^2+b^2}}\right) + \arccos(\frac{a}{\sqrt{a^2+b^2}})$$

$$\beta = \arccos(\frac{L_4\cos\varphi + L_3\cos\theta - L_3\cos\gamma}{L_2})$$

The intermediate variables are *a*, *b* and *c*:

$$a = 2L_1 \cdot (L_3\sin\theta - L_4\sin\phi)$$

$$b = -2L_1 \cdot (L_3\cos\theta + L_4\cos\phi)$$

$$c = L_2^2 - L_1^2 - (L_3\cos\theta + L_4\cos\phi)^2 - (L_3\sin\theta - L_4\sin\phi)^2$$

With the inclination angle δ of the top beam, the top beam and the base structure are analyzed respectively, and the coordinates of the front column pin point C, the rear column pin point D, the front column pin point A and the rear column pin point B in the base coordinate system (the rear connecting rod pin point is the origin) can be determined.

In this way, the angle η between the front column and the base, the length L of the column in this process can be calculated.

$$\eta = -\arcsin\frac{Y_{\mathrm{AC}}}{X_{\mathrm{AC}}} \tag{8.5}$$

$$L_{length} = \sqrt{{X_{\mathrm{AC}}}^2 + {Y_{\mathrm{AC}}}^2} - L_{\mathrm{ACoriginal}} \tag{8.6}$$

8.3.3.2 Calculation of Actual Rotation Angle and Parameters

In the actual working process of the hydraulic support, there are lateral and longitudinal tilt angles in the bottom plate, which may lead to the deflection of the hydraulic support at any time. In the actual working process, the attitude of hydraulic support must fully consider the transverse and longitudinal inclination.

According to the principle of inclination sensor, when there is an angle between the plane where the sensor measures the angle and the vertical plane, as shown in Fig. 8.7, the measured angle will no longer be the angle in the calculation plane, so it needs to be converted to improve the accuracy of the measurement data.

According to the values and actual conditions of four biaxial inclination sensors, when the hydraulic support base has angle with horizontal and vertical plane, then the two angles shall be converted into the angle in the calculation plane at the same time. The relationship between the sensor value and the calculated angle is shown in Table 8.3.

Setting the height of hydraulic support is a key parameter of hydraulic support. According to the angle conversion results, the calculation formula can be obtained.

$$H = f(\alpha_1', \alpha_2', (\alpha_3')\ or\ (\alpha_4')) \tag{8.7}$$

8.4 Seamless Linkage Method of Hydraulic Support Components Based on Unity3D

8.4.1 Overall Idea of Virtual Simulation Method for Hydraulic Support

In order to build a real virtual model of hydraulic support in VR environment, we must have enough understanding of its structure performance and VR simulation environment. Now 3D modeling mainly depends on CAD software. The differences,

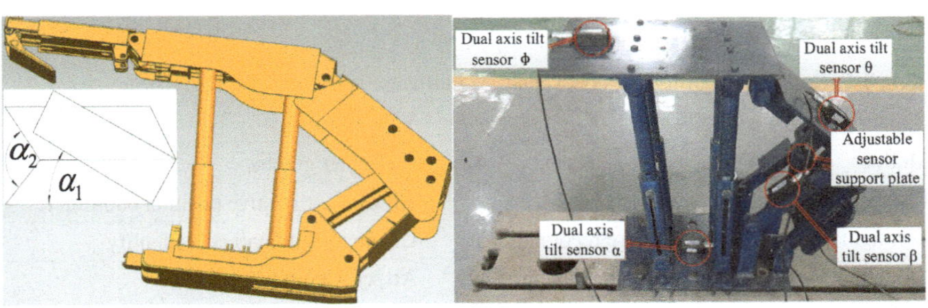

Fig. 8.7 Support attitude diagram under actual working condition

Table 8.3 Relationship between sensor value and calculated angle

Name	Significance	The sensor actually measures the angle	Calculation angle
X_1	The angle between the base tilt sensor and the horizontal plane	α_1	$\alpha_1' = \alpha_1$
Y_1	The angle between the base tilt sensor and the vertical plane	β_1	$\beta_1' = \frac{\beta_1+\beta_2}{2}$
X_2	The angle between the top beam inclination sensor and the horizontal plane	α_2	$\alpha_2' = \arcsin\frac{\sin(\alpha_2-\alpha_1)}{\cos\beta_1}$
Y_2	The angle between the top beam inclination sensor and the vertical plane	β_2	$\beta_1' = \frac{\beta_1+\beta_2}{2}$
X_3	The angle between the rear linkage angle sensor and the horizontal plane	α_3	$\alpha_3' = \arcsin\frac{\sin(\alpha_3-\alpha_1)}{\cos\beta_1}$
X_4	The angle between the tilt sensor of shield beam and the horizontal plane	α_4	$\alpha_4' = \arcsin\frac{\sin(\alpha_4-\alpha_1)}{\cos\beta_1}$

advantages and disadvantages of CAD software and VR software are fully analyzed, and their characteristics and advantages are complemented and integrated, in order to show the real characteristics of the stent in the virtual environment. Table 8.4 shows the differences in configuration and expression between CAD software and VR software.

CAD software includes UG, Pro/E, CATIA and SolidWorks. This section takes UG as an example for analysis. The STL format is imported into 3Dmax software to repair and add materials to the model, and then converted into an intermediate format that can be imported into VR software. After the intermediate format is exported from CAD, the position information of the model is saved, but all the mechanical constraints are lost. After many experiments, a set of input rules is found. If the model of each part is imported into VR software according to this rule, the relative position of each part will be the same as the correct assembly format in CAD software. Using this feature, in CAD software, the key feature points of each part movement are

Table 8.4 Comparison of features between CAD software and VR software

Name	CAD software	VR software
Motion	Follow the constraints	Computer graphics
Location	Follow the constraints	Absolute or relative position coordinates
Distinguish	Automatic feature recognition	Unable to automatically identify
Modelling	Modeling of mechanical products	Art, curve modeling
Display	No material	Rich materials

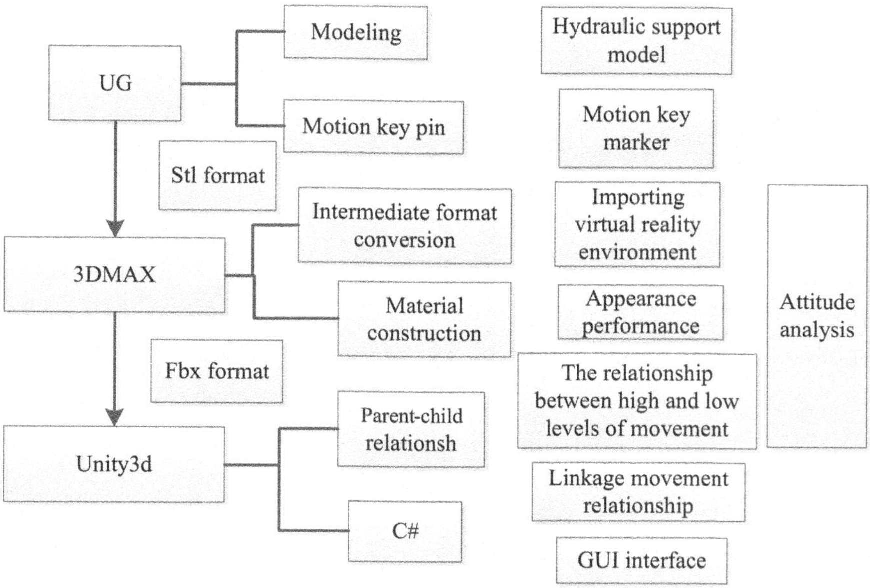

Fig. 8.8 Sequence flow chart of seamless linkage method

assembled with a specific marker and used to mark the key feature points, which can realize and ensure the authenticity and reliability of the part movement information.

After importing into VR environment, the parent-child relationship between components is established (The movement of the parent affects the movement of the child, and the movement of the child does not affect the movement of the parent). The attitude of each component is analyzed, and the relationship between each motion state variable is solved to avoid redundancy. Using C# to write programs to achieve real movement speed, state switching relationship and the establishment of GUI interface, so as to complete the virtual linkage. Figure 8.8 is the flow chart of seamless linkage method of hydraulic support components in this section.

8.4.2 Model Construction and Repair

According to the full set of drawings of ZZ4000/18/38 hydraulic support, modeling and model repair are carried out in UG software. As shown in Fig. 8.9, for the rotation center point of the motion relationship, establishing the pin axis, importing each part into 3Dmax in STL format, and then exporting the model in FBX format. At this time, the model can be imported into Unity3D software, and the position relationships between the repaired 11 pin shafts and all component models are consistent with that of the parts repaired in UG, so as to mark the motion center point.

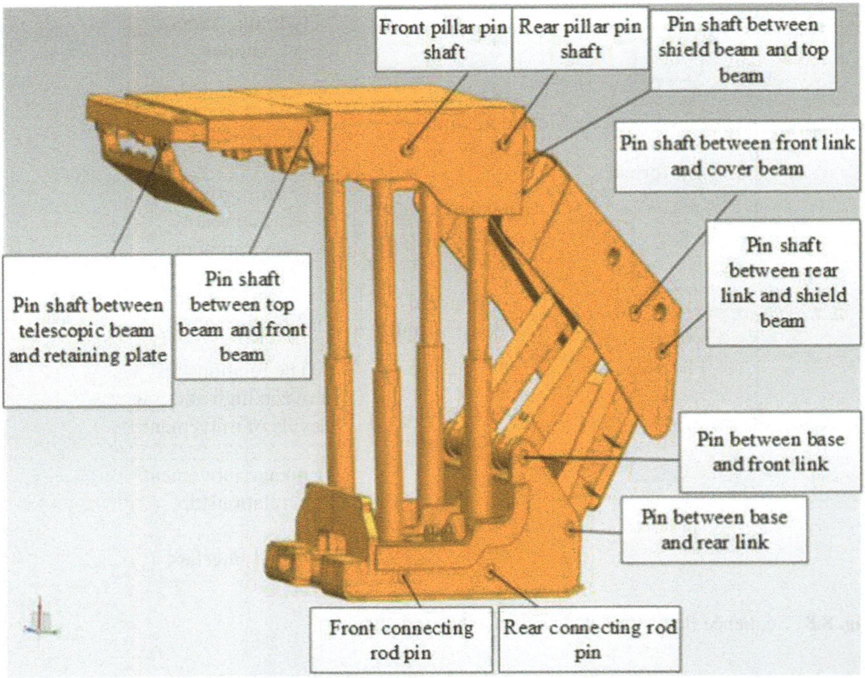

Fig. 8.9 Model of hydraulic support repaired in UG

In the process of format conversion, because the local coordinate system of the front and rear column model in UG does not coincide with the overall coordinate system of the hydraulic support, it is impossible to accurately find out the coordinate axis of the linear movement of the cylinder rod relative to the cylinder body in 3DMAX, which leads to the deviation of the cylinder rod movement in Unity3D. After many tests, the STL file was exported from the cylinder sub model assembly in UG and the position was corrected in 3DMAX.

8.4.3 Seamless Linkage Method Between Virtual and Reality

After importing into Unity3D, we need to use the following key technologies:

8.4.3.1 Establishment of Parent-Child Relationship

To establish a parent-child relationship, first, you need to establish a hierarchical relationship in the hierarchy view, as shown in Fig. 8.10. Then you need to establish a hierarchical relationship ZzyyControl.cs script, assign to the base object, define

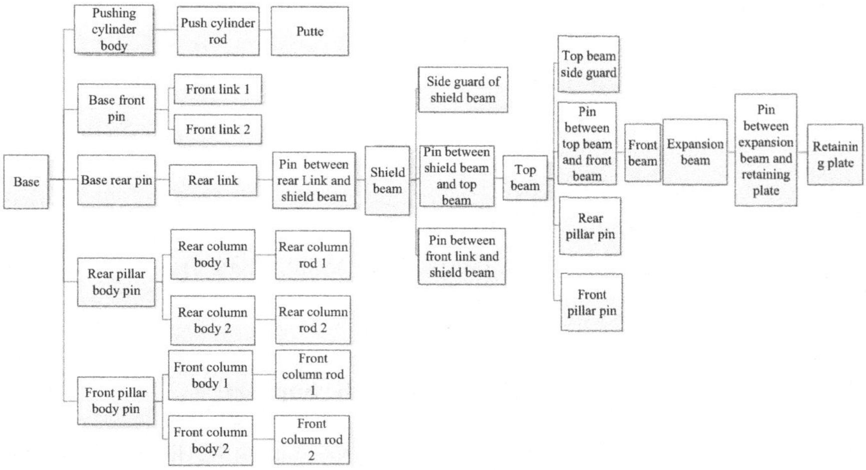

Fig. 8.10 Structure diagram of parent-child relationship of hydraulic support components

the component variables by C# and establish a one-to-one correspondence between these variables and the component.

After variable definition, using "GameObject = gameobject1.transform.GetChild (). transform;" to establish the parent-child relationship between components.

8.4.3.2 Establishment of Local Coordinate System and Global Coordinate System

Since the relative positions of all parts with the base remain unchanged, the local coordinate system must be used for motion analysis.

The localPosition function is used for the movement of the push cylinder, and the coordinates before the movement are (0.12, −6.41, 0).

TuiYiYouGangShenChang is the extension variable of push cylinder, which is realized with the following code:

$$\text{TuiYiYouGangGan.localPosition} = \text{new Vector3}(0.12\text{f}, \\ -6.41\text{f} - \text{TuiYiYouGangShenChang}, 0);$$

The pin between expansion beam and retaining plate controls the rotating plate's rotational motion. Using the localRotation function, quaternion is used to represent the rotation in Unity3D:

$$\text{Quaternion} = (\text{xi} + \text{yj} + \text{zk} + \text{w}) = (\text{x}, \text{y}, \text{z}, \text{w})$$

$$\text{Q} = \cos(\text{a}/2) + \text{i}\,(\text{x} \cdot \sin(\text{a}/2)) + \text{j}(\text{y} \cdot \sin(\text{a}/2)) + \text{k}(\text{z} \cdot \sin(\text{a}/2))$$

(a is the rotation angle);

Using the following code to achieve:

$$\begin{aligned}&\text{HuBangBanXiaoZhou.localRotation}\\&\quad = \text{new Quaternion}\\&\quad\ (0, 0, \text{Mathf.Sin(HuBangBan JiaoDu} * \text{Mathf.PI}/360),\\&\quad\ \text{Mathf.Cos(HuBang BanJiaoDu} * \text{Mathf.PI}/360))\end{aligned}$$

8.4.3.3 Realization of Each Action of Hydraulic Support

Combined with the Finite State Machine (FMS), the state of the hydraulic support {rolling out the scraper-trough conveyer (0), taking back the face guard (1), descending column (2), moving (3), ascending column (4), extending the face guard (5)} is established.

The state switching condition is shown in Fig. 8.11. The "switch… case…" statement achieves several different states with the change of shearer position and automatically switch their running state. Among them, m, n and q are the parameters of the taking back the face guard, the moving process and the rolling out the scraper-trough conveyer respectively.

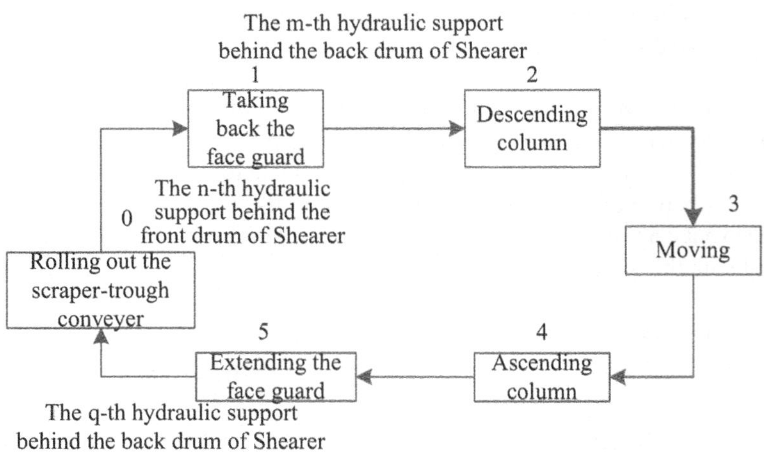

Fig. 8.11 Finite state machine model of hydraulic support

8.4.3.4 VR Velocity Solution

Taking the movement of ZZ4000/18/38 hydraulic support driven by XR-WS1000 emulsion tank as an example, the velocity solution is analyzed. The basic parameters of the emulsion tank are: nominal pressure 31.5 MPa, nominal capacity 1000 L, nominal flow 125 L/min. According to the ideal state, taking the column as an example, the movement speed of the cylinder is calculated.

The diameter of the non-rod cavity is 200 mm, the diameter of the rod cavity is 85 mm, the total number of front and rear columns is 4, and the height of the falling column is assumed to be 200 mm.

The velocity of the non-rod cavity is:

$$\begin{aligned} \mathrm{V1} &= \mathrm{Q1/A} = 125 \times 103/((3.14 \times 0.12 \times 104) \times 4) \\ &= 99.52(\mathrm{cm/min}) = 1.66\,\mathrm{cm/s} \end{aligned}$$

$$\text{The elongation time}\, 200\,\mathrm{mm}/16.6\,\mathrm{mm/s} \times 1.2 = 14.46\,\mathrm{s}$$

Unity3D software can set the number of frames refreshed per second. Changing the option "V Sync Count" to the option "Don't Sync", and then adding the UpdateFrame.cs script to modify the frame number:

```
Application.targetFrameRate = target FrameRate;
```

targetFramerate = 10 means that the program executes 10 frames a second, and the corresponding update() function executes 10 times.

Assuming that the top beam rises 200 mm in the process of column rising, the corresponding rear column rises 201.834 mm according to the front pose calculation results. According to the calculation, the process time is 14.46 s, so the increment of each frame is 201.834/(14.46 × 10) = 1.39 mm. The inclination angle of the rear column is changed from 86.8° to 86.6°, and then the cylinder is extended through the following cycles:

```
if (DiZuoHouLiZhuShenChang < 201.834f)
{
.....loop statement
DiZuoHouLiZhuShenChang +=1.39f;
}
```

8.4.3.5 Rotation Angle of Top Beam Offsetting Shield Beam

As the sub object of the shield beam, the top beam will rotate with the rotation of the shield beam, so the angle compensation of the top beam in the opposite direction of the rotation direction of the shield beam should be carried out to ensure the correct attitude of the top beam. The inclination angle of the top beam is driven by the inclination variable of the top beam independently, and the influence of the angle change of the cover beam is eliminated during the action of the cover beam. It is implemented by the following code:

```
YanHuDingLiangXiaoZhou.localRotation
  = new Quaternion(0, 0, Mathf.Sin((Ding LiangRotAngle
    − YanHuLiangQing JiaoAngle) * Mathf.PI/360),
  Mathf.Cos((DingLiangRotAngle
    − YanHuLiangQing JiaoAngle) * Mathf.PI/360));
```

8.4.3.6 The Transformation of the Relationship Between Parent-Child Relationship in the Process of Moving and Pushing

In the process of moving the frame, the top beam is separated from the top plate, and the hydraulic support is pulled with the scraper conveyor as the fulcrum. When pushing and sliding, the top beam is in close contact with the top plate, and the scraper conveyor is pushed with the hydraulic support as the fulcrum. In the VR environment, the sliding cylinder doesn't move with the bracket during the frame moving process, so after the sliding is completed, the parent-child relationship between the sliding cylinder rod and the sliding cylinder body must be temporarily separated, which is achieved by the following code:

```
TuiYiYouGangTi.transform.DetachChildren();
```

After moving the frame, setting the parent object of the pushing rod as the pushing cylinder body, and moving with the parent object again:

```
TuiYiYouGangGan.transform.parent = TuiYiYouGangTi;
```

8.4.3.7 The Attitude of the Hydraulic Support in Actual Working Conditions

In the actual working process of the hydraulic support, there are lateral and longitudinal inclination angles of the bottom plate, as well as the crooked frame. The pitch angle, roll angle and yaw angle of the shearer are defined as the pitch angle, roll angle and yaw angle of the fully mechanized face. The mutual change of three angles shows the change of the hydraulic support in the actual working process, and then also shows the change of the topographical conditions of the fully mechanized mining face, which is realized by the following code:

```
transform.eulerAngles = new Vector3(RollAngle, YawAngle, PitchAngle);
```

8.4.3.8 GUI Interface

Using Unity3D software's own UI to design and setting the position and movement direction of the shearer. The random function is used to realize the change of the position and movement direction of the shearer, and the support will perform corresponding actions with the change of the shearer. Establish the control buttons for the column, push and slide, move the frame and the guard plate respectively, and the user can carry out remote manual intervention operations according to these buttons.

8.4.4 Human-Computer Interaction Modes and Methods

The development of the hydraulic support virtual simulation system is based on the technology of Sect. 8.4.3, and the system has two interactive modes. The first interactive mode is the virtual GUI (Graphical User Interface) interactive mode, refer to Sect. 8.4.3.8. The second way of interaction is virtual hand interaction, which uses 5DT data glove and position tracker Patriot to complete the experiment. In the Unity3D software environment, the established operation valve model is grasped. The data glove controls the posture and data of the joints of the virtual hand, the position tracker determines the position of the virtual hand. When the ray of the finger intersects the surrounding ball of the operating handle, it is in contact with the object. According to the actual grasping law of the human hand, when the three fingers including the thumb touch the operating handle and all bend below 2/3 of the entire palm plane, it is judged that the handle has been grasped. Keep the hand held, you can continue to operate the handle to hit the left or right position. The following Table 8.5 is the established operating handle function table:

Table 8.5 Function list of hydraulic support handle

Operating handle position	Left position	Right position
First row	Column raised	Column descended
Second row	Frame moved	Scraper-trough conveyer push
Third row	Front beam extended	Front beam retracted
Fourth row	Side guards extended	Side guard retracted
Fifth row	Telescopic beam extended	Telescopic beam retracted
Sixth row	Guard board extended	Guard board retracted

The release rule of the virtual hand is opposite to the grasping rule. When three or more rays are far away from the surrounding ball and don't intersect with it, it is judged that the virtual hand releases the operating handle.

To test the virtual simulation system of hydraulic support, first click GUI button with mouse:

- After entering the system, we continue to press the "+" next to the "Push scraper-trough conveyer" button, and the system will gradually extend the push slide bar. When the push rod is in place, we will see the prompt of "In place" in the system. At this time, we can release the button;
- When we click the "+" of the "Column" button, the system will appear the action of column rising, and the prompt of "Reached the highest point" will appear after the column rises to the highest point. In the process of movement, we carefully observe the movement relationship of each component, and the system can achieve mutual coordination and seamless linkage;
- When we click "Move bracket", the system will move the hydraulic support, and its speed is consistent with the actual support speed, as shown in Fig. 8.12.

Then the system performs a virtual hand operation virtual handle test in the VR laboratory, as shown in Fig. 8.13 for the virtual hand test chart. The virtual hand grabs the second row of operating handles and hits it to the left position. The virtual

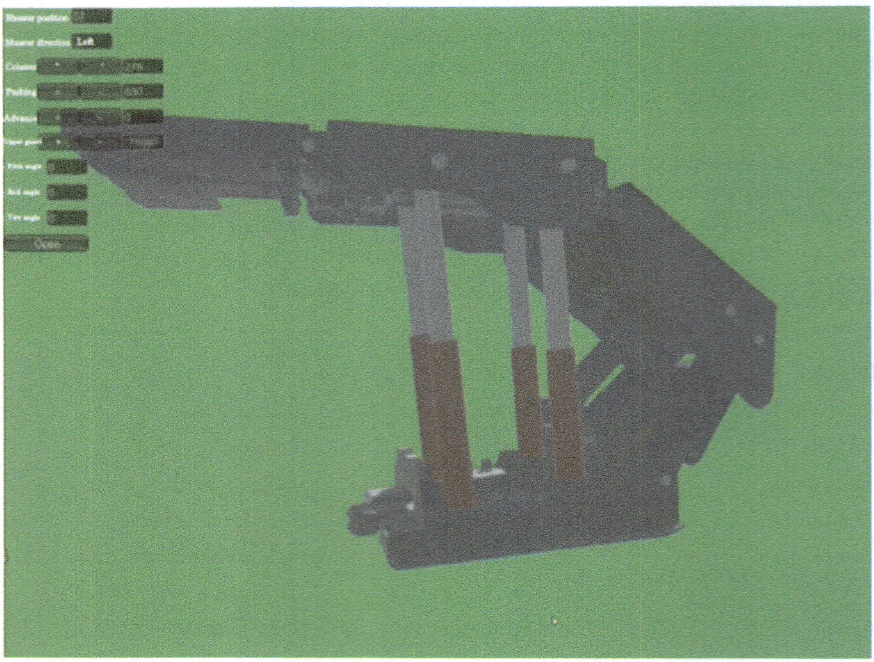

Fig. 8.12 Moving state of hydraulic support in virtual picture

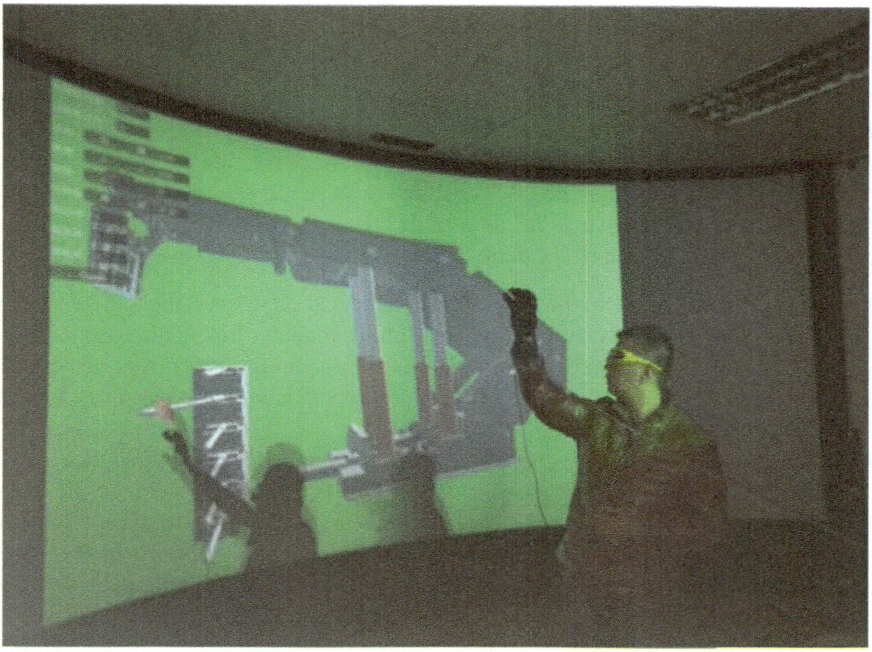

Fig. 8.13 VR laboratory test

hydraulic support will move the frame. After the frame-moving distance reaches the step distance, the system will display the prompt "Move the frame in place".

The test results of the whole system are good, and the various motion states of the hydraulic support can be truly reproduced, and the various components are seamlessly linked during the operation of various actions. Through careful analysis, the system does not appear to be mismatched and uncoordinated.

8.5 Virtual Bending Technology of Scraper Conveyor Based on Unity3D

In a fully mechanized mining face, the scraper conveyor is coupled with the bottom plate in real time, which will show the fluctuation of the bottom plate. During the cutting process of the shearer, the scraper conveyor will also have an "S-shaped" curved section. Each section of the middle trough may be bent at a certain angle in two directions with the adjacent middle trough. How to perform virtual bending of the scraper conveyor is also a key technology.

8.5.1 Model Construction and Repair

Take the chain-link type scraper conveyor as an example for analysis. The repair effect of the middle trough n is shown in Fig. 8.14, where n is the serial number of the middle trough. The front pin shaft n, the rear pin shaft n, the pin row pin shaft n left and the pin row pin shaft n right are added respectively. To mark the function of the longitudinal and transverse bending of the central trough. In Fig. 8.14, the front pin n + 1 and the rear pin n + 1 correspond to the middle trough n + 1:

- Rotate in the Z direction: the front pin n and the rear pin n mark the middle trough n. During the bending process, the front link or the later link is the base point to rotate around the Z direction axis, the front pin n and the rear pin n. The center point in the Y direction is the axis of rotation, which is mainly used to advance to the side of the coal wall;
- Rotate in the Y direction: the front pin shaft n and the rear pin shaft n mark the middle trough n and rotate around the Y axis as the base point of the front link or the later link during the bending process; mainly used when the bottom plate is uneven, passing. The calculation of the coupling relationship between the bottom of the working face and the middle trough can be self-adaptively laid on the bottom;
- Pin row pin shaft n left and pin row pin shaft n right are used to mark the walking path of the shearer, respectively in the two pin row seat holes of the middle trough n.

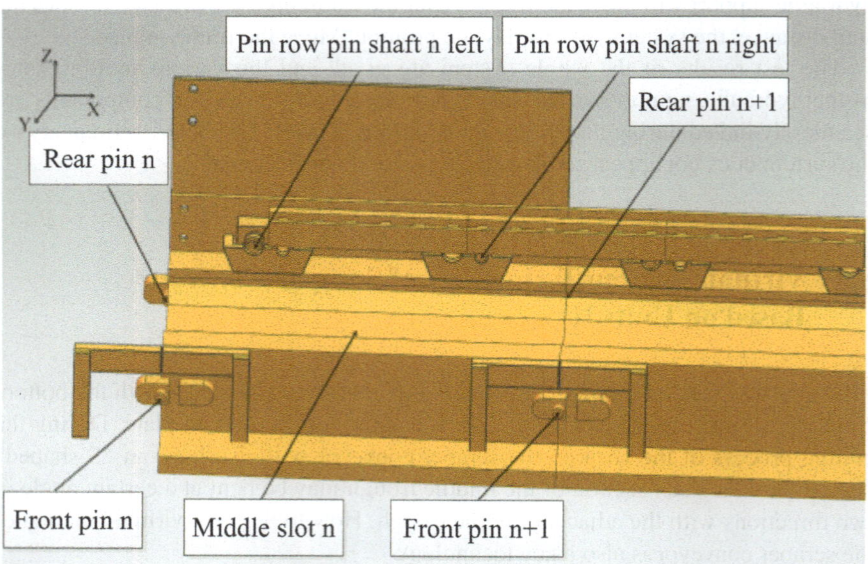

Fig. 8.14 The repaired middle trough model

8.5.2 *Virtual Bending Technology of Scraper Conveyor*

8.5.2.1 Analysis of the Formation Process of the Bending Section and Solution of Bending Parameters

The chutes of the scraper conveyor are connected in the form of dumbbell pins or collars, and each section of the chute is connected to the hydraulic support through a pushing cylinder. When the front and rear drums of the shearer finish cutting the front coal seam, the pushing cylinder of the hydraulic support will extend and push the chute toward the goaf, and each chute will be bent under the connection of dumbbell pins or collars. With the different extension lengths of the hydraulic supports, each chute can form two symmetrical curved sections with equal lengths and opposite directions, referred to as "S-shaped".

The shearer also has to pass through such an "S-shaped" curved section when it cuts into the knife diagonally to achieve the effect of advancing a deep cut.

In the "*Continuous Conveying Machinery Design Manual*" and Jiang Xueyun's detailed calculation method for the length of the bending section, the latter formula is more explanatory. The cutting depth of the shearer in this laboratory is 630 mm, and the number of central troughs on a single side is N = 5, so there are 2N − 1 = 9 sections in the bending section. Inversely, the accurate value of the bending angle α_a of adjacent central troughs is 0.967°, which is approximately 1°.

8.5.2.2 Virtual Bending and Transfer Process

Figure 8.15 shows the process of activating the moving process of the hydraulic support one by one as the position of the shearer changes during the operation of the shearer, thereby gradually forming a bending section.

8.6 Virtual Memory Cutting Method of Coal Shearer Based on Unity3D

The simulation of the shearer mainly includes the virtual height adjustment and virtual walking of the left and right rocker arms and the left and right height adjustment cylinders. It is relatively easy to implement in Unity3D software, so this section will not specifically introduce the virtual simulation of the shearer attitude. The shearer needs to be coupled with the shape of the scraper conveyor in real time when it is walking. Now the shearer memory cutting technology has been widely used and has achieved positive results. In order to cut the roof and floor curves, the virtual shearer needs to have the same memory cutting function as the real shearer.

VR technology can construct a real 3D scene of a fully mechanized mining face similar to a panoramic view, simulating the operation process and the operation of

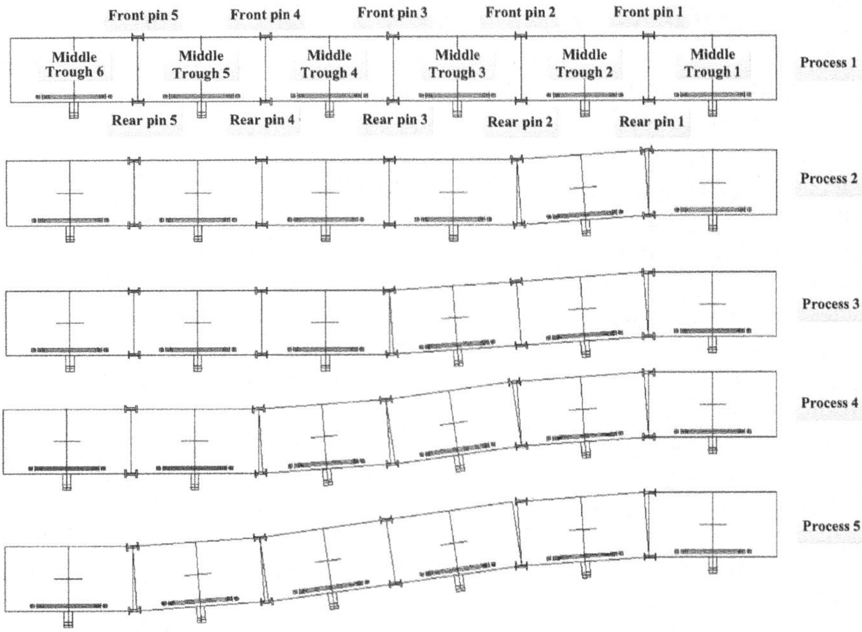

Fig. 8.15 The initial parent-child relationship diagram when the scraper conveyor doesn't begin to bend

process equipment. If it is possible to establish a digital virtual model of the shearer that is in line with the reality and a shearer information model that can reflect the operating characteristics of the shearer, the two will be integrated with each other and the operation mode will follow the memory cutting "Teaching-Execution-Correction-Execution". Then the shearer's memory cutting technology can be quickly tested in the VR environment.

8.6.1 *Theories and Methods of Virtual Memory Cutting*

The shearer memory cutting is the memory of the key information of the shearer in a cycle cutting process, and based on the key information recorded, the cutting state and information in the memory process can be reproduced. At the same time, memory cutting can highly compensate the height of the front and rear drums in real time according to the real-time changes of the terrain trend.

Therefore, the establishment of a virtual memory cutting method needs to solve the following problems:

- Setting of terrain floor and shearer walking conditions;
- How to make the virtual shearer record the virtual information in the same way as the real onboard controller records the status information;
- During the next cycle of operation of the virtual shearer, the corresponding cutting operation data is automatically generated according to the recorded information and the selected single-blade and double-blade memory cutting method. At the same time, the virtual shearer can be highly compensated according to real-time terrain changes during operation.

Therefore, the shearer virtual memory cutting method is divided into two steps. First, the virtual roof and floor environment is generated in the Unity3D environment, and then the virtual scraper conveyor is laid on the virtual floor as the track on which the virtual shearer runs. In the demonstration stage, click the virtual manipulation button to operate the virtual shearer, and the virtual controller stores, analyzes and processes the operating data in real time. In the execution phase, the information recorded and stored by the virtual controller is used to reproduce the cutting state. According to the information of the virtual sensor, the virtual shearer compensates the cutting height in real time. The overall research idea is shown in Fig. 8.16.

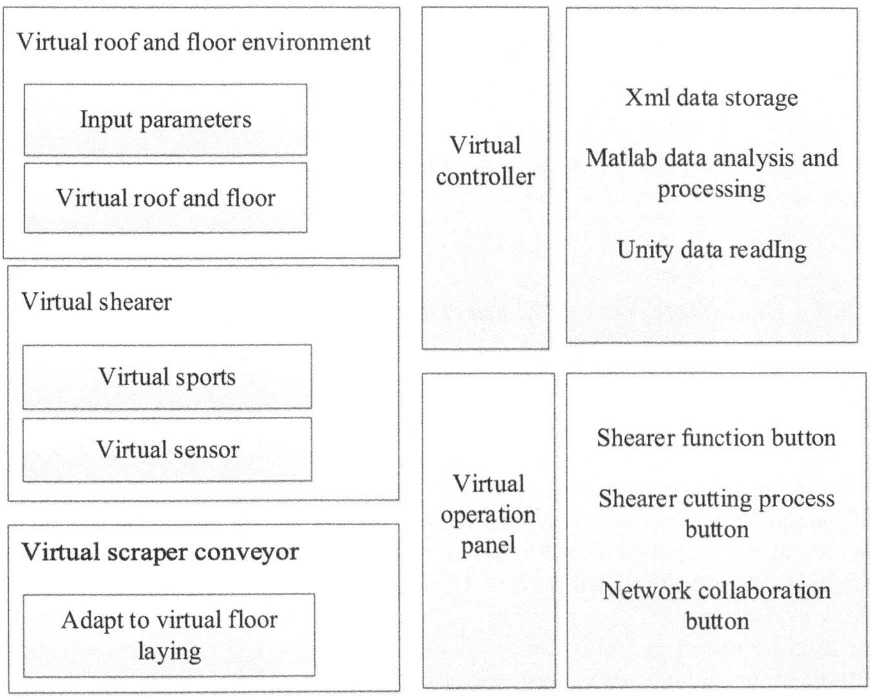

Fig. 8.16 The overall research idea of the virtual memory cutting method

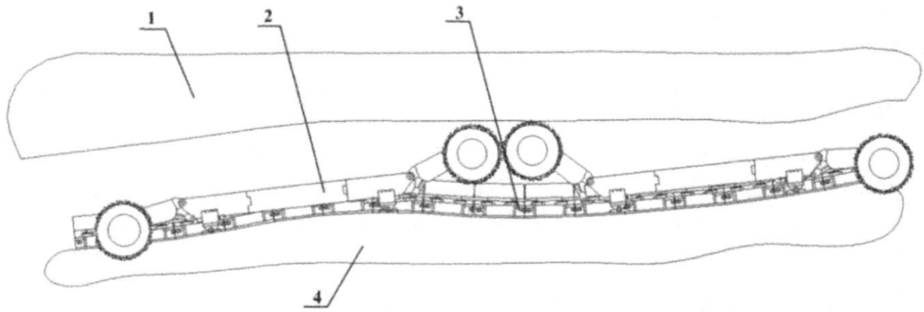

Fig. 8.17 One-way demonstration memory cut solution. 1—Top plate; 2—Shearer; 3—Scraper conveyor; 4—Bottom plate

8.6.2 Mathematical Model of Memory Cutting

Memory cutting is mainly divided into two-way demonstration memory cutting and one-way demonstration memory cutting. The former is mainly for the reproduction of recorded information, and the latter needs to solve the cutting information in the "execution" mode based on the information of a single demonstration walking.

Memory cutting requires repeated use of various cutting information that has been stored, including the length of the working face, the traction direction, the traction speed, the position of the left and right drums, the lateral inclination of the shearer, the longitudinal inclination of the shearer, and the location of the shearer. According to the mathematical model of one-way memory cutting, it is compiled and solved in Unity3D, as shown in Fig. 8.17.

8.6.3 Real-Time Virtual Shearer Drum Height Compensation Strategy

An important part of the shearer's memory cutting is the good control of the drum height. When the bottom plate is at different angles, the drum needs to be adjusted to a suitable height to ensure the normal cutting work of the shearer. The topography changes slightly after each cut by the shearer. Therefore, the shearer needs to obtain the lateral and longitudinal inclination in real time, and then compensate the drum height in real time. As shown in Fig. 8.18, the sensor measures the pitch or roll angle of the fuselage, and uses the geometric relationship of a series of parameters such as the fuselage rocker arm to calculate the relationship between the inclination angle and the drum height when the fuselage is at different inclination angles.

The cutting heights $H1$ and $H2$ of the shearer under the conditions of uphill and downhill are:

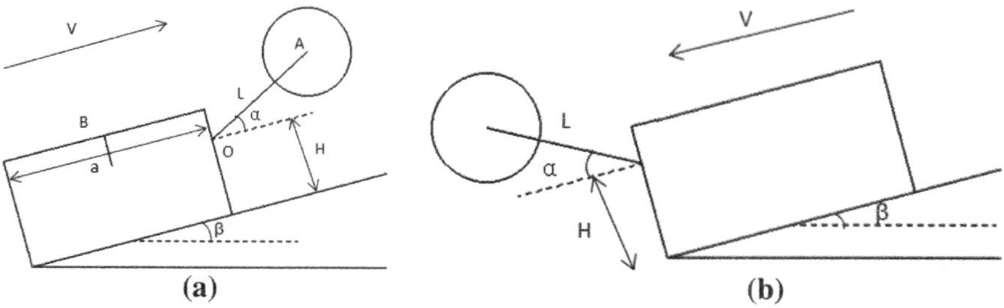

Fig. 8.18 The relationship between the shearer up and down drum and the inclination angle of the fuselage. **a** Schematic diagram of the shearer uphill; **b** Schematic diagram of the shearer downhill

$$H_1 = \frac{H + L \cdot \sin\alpha}{\cos\beta} + (L \cdot \cos\alpha + a) \cdot \sin\beta - (H + L \cdot \sin\alpha) \cdot \tan\beta \sin\beta + R \tag{8.8}$$

$$H_2 = \frac{H + L \cdot \sin\alpha}{\cos\beta} - L \cdot \cos\alpha \sin\beta + (H + L \cdot \sin\alpha) \cdot \tan\beta \sin\beta + R \tag{8.9}$$

Figure 8.19 is the relationship between the inclination angle of the drum and the fuselage during the upward and downward mining of the shearer. Where y is the width of the drum, b is the width of the shearer, R is the radius of the drum, γ is the angle of inclination, and OB is the projection of the rocker arm projection on the plane.

The cutting height H_3 of the upward mining is:

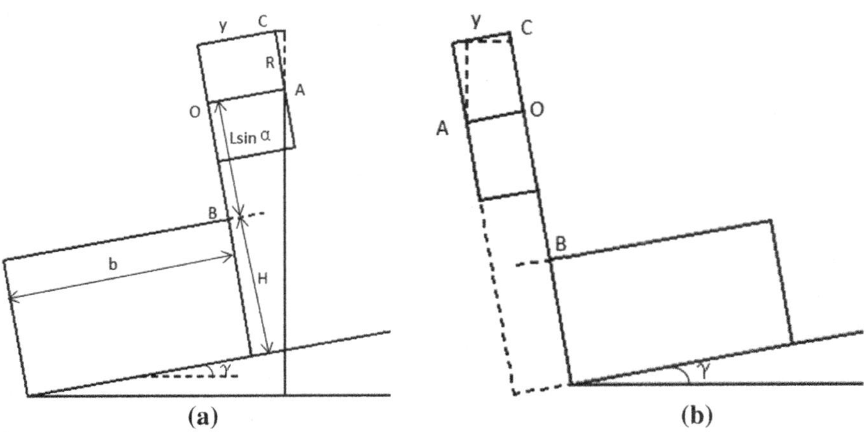

Fig. 8.19 The relationship between the shearer's pitch mining drum and the inclination angle of the fuselage. **a** Schematic diagram of shearer over mining. **b** Schematic diagram of shearer under mining

$$H_3 = \frac{H + L \cdot \sin\alpha}{\cos\gamma} + (b + y)\sin\gamma - (H + L \cdot \sin\alpha) \cdot \tan\gamma \sin\gamma + R \cdot \cos\gamma + (H + L \cdot \sin\alpha) \cdot \tan\gamma \sin\gamma + R \cdot \cos\gamma \quad (8.10)$$

The cutting height H_4 of the down mining is:

$$H_4 = \left[H + L\sin\alpha - y \cdot \tan\gamma\right] \cdot \cos\gamma + \frac{R}{\cos\gamma} + (y - R \cdot \tan\gamma) \cdot \sin\gamma \quad (8.11)$$

8.6.4 Virtual Controller

The virtual controller can mainly complete data storage, data analysis and processing, and data reading functions. The overall function is shown in Fig. 8.20.

Among them: the data storage module collects data from the installed virtual sensor according to requirements and stores it in an xml file during the operation of

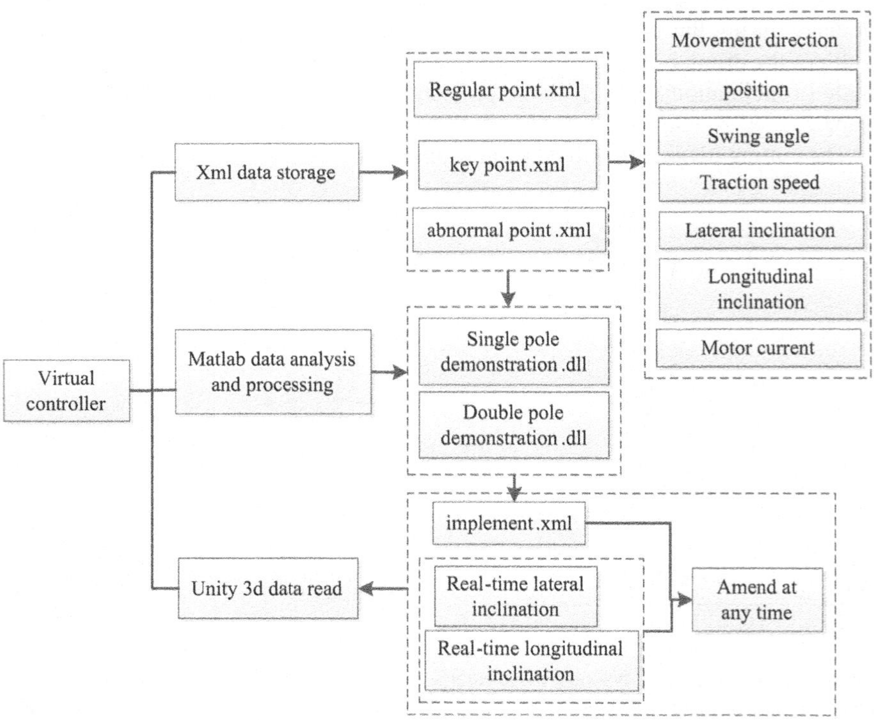

Fig. 8.20 Overall functions of the virtual controller

the virtual shearer. The data is divided into three categories, namely regular points, key points and abnormal points.

The regular point is a set of virtual sensor data automatically collected after the shearer walks the length of a middle slot and stored in the changgui.xml file.

The key point is the point where the shearer receives virtual control instructions and changes its working posture. In this system, the virtual sensor data at the two moments when the function button of the virtual shearer operation panel is pressed and released is stored in the guanjian.xml file.

The abnormal point refers to the virtual sensor data at the moment when the current of the virtual shearer motor is abnormal when the height of the left or right drum of the shearer exceeds the range of the virtual roof or the virtual floor. Or it is the virtual sensor data at the moment when the height of the left or right roller returns to the range of the virtual top plate or the virtual bottom plate after manual intervention. The abnormal point is stored in the yichang.xml file.

The data analysis and processing module compiles the dll file in the matlab software to integrate the data in the three xml files (changgui.xml, guanjian.xml and yichang.xml) collected by the data storage module. It uses a specific algorithm to integrate and generate the next cut data, and generates virtual sensor data that can be executed in the memory cutting mode in the.xml file. For example, if you select single pole demonstration and double pole demonstration respectively, the system will generate single pole demonstration.dll file and double pole demonstration.dll file respectively, which can be realized by Get_jyjg() function.

```
public void Get_jyjg(int jyjg_ID, float jyjg_Position, bool jyjg_Direction,
float jyjg_Speed, float jyjg_ZuoRotAngle, float jyjg_YouRotAngle, float
jyjg_HengXiangAngle, float jyjg_ZongXiangAngle, bool jyjg_DiJiDianLiu) { }
```

The data structure of changgui.xml is as follows:

```
<ROOT>
<jyjg jyjgID="1" jyjgPosition="43.6856" jyjg_Direction="True"
jyjgSpeed="10" jyjgZuoRotAngle="30" jyjg_HengXiangAngle="5"
jyjg_ZongXiangAngle="2" jyjg_DianJiDianLiu=true/>
</ROOT>
```

The meaning of each variable is shown in Table 8.6.

The data structure of guanjian.xml and yichang.xml is the same as that of changgui.xml. The three xml files are used to uniquely identify process points through jyjgPosition and jyjg_Direction.

The data reading module refers to reading the corresponding virtual sensor data of the shearer in the execution.xml generated by the data analysis and processing module when the virtual shearer runs to the corresponding position in the memory mode, and cuts out the artificial demonstration shape. In this mode, the shearer compensates for the height of the drum as the lateral and longitudinal inclination angles change.

The specific process is as follows: the virtual coal mining machine reads the previous several values in advance and performs interpolation and other operations. The fixedupdate() function is used to detect the position and direction conditions in

Table 8.6 Record structure variables of the shearer xml file

Number	Variable name	Meaning
1	jyjg jyjgID	Information point ID
2	jyjgPosition	Virtual shearer location
3	jyjg_Direction	Traction direction of virtual shearer (true for left, false for right)
4	jyjgSpeed	Traction speed of virtual shearer
5	jyjgZuoRotAngle	Left rocker angle of virtual shearer
6	jyjgYouRotAngle	Right rocker angle of virtual shearer
7	jyjg_HengXiangAngle	Lateral inclination of virtual shearer
8	jyjg_ZongXiangAngle	Longitudinal inclination of virtual shearer
9	jyjg_DianJiDianLiu	Virtual shearer motor current (true means normal, drum doesn't exceed the range of virtual roof or virtual floor, false means abnormal, drum exceeds the range of virtual roof or virtual floor)

each frame. If the conditions are met, the position information points at this time are given the attitude information of the virtual shearer, so as to drive the Shearer to move according to the pre trajectory. It is mainly realized by Loadxml() function:

```
Void public LoadXml();
```

8.6.5 *Virtual Interaction*

The virtual interaction mainly completes the operation functions of the virtual shearer. The function classification of the virtual operation panel is shown in Fig. 8.21. The operator can use the GUI buttons to adjust the front and rear drums of the shearer according to the generated roof and floor curves. During operation, the virtual controller will record relevant memory cutting key data in real time.

8.6.6 *Shearer Virtual Memory Cutting Interface*

After the development is completed, the shearer virtual memory cutting interface is shown in Fig. 8.22.

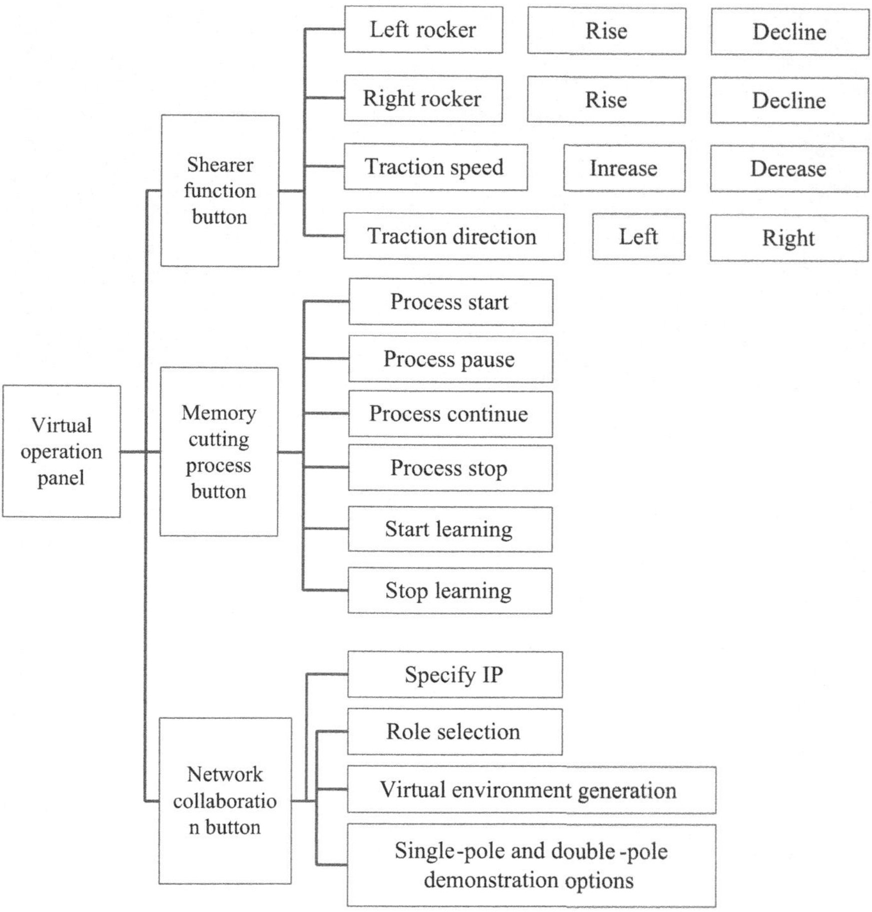

Fig. 8.21 Classification of virtual operation panel

8.7 Summary

This chapter studies the working condition monitoring and virtual simulation method for the single machine in the "three machines" of the fully mechanized mining face. In the working condition monitoring part, the monitoring method of actual working condition attitude is studied for the coal shearer, scraper conveyor and hydraulic support stand-alone. Using the established physical information sensing system, a single machine working condition monitoring method was found. The single-machine virtual simulation method is to study the virtual "three-machine" single-machine virtual simulation method that is completely consistent with the actual "three-machine" in the Unity3D environment. Including posture analysis, model construction and repair and compilation of virtual behaviors, seamless linkage

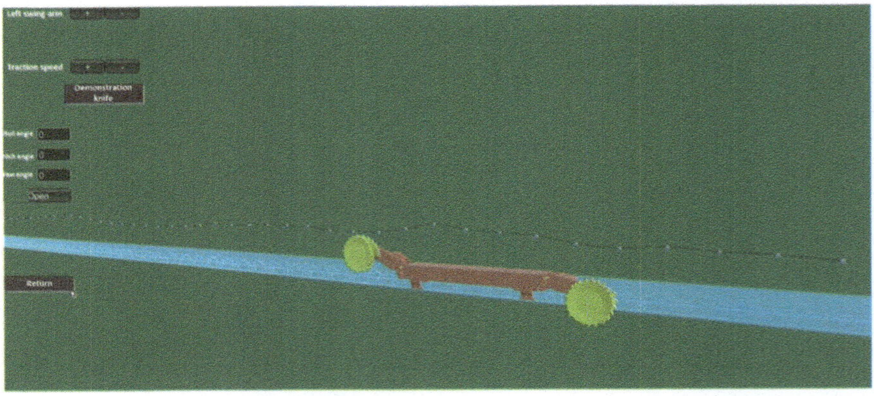

Fig. 8.22 Virtual memory cutting interface

method of hydraulic support components and virtual hand-man–machine interaction mode, virtual bending technology of scraper conveyor and virtual memory cutting method of shearer.

Chapter 9
The Method and Technology of Virtual Collaborative Simulation Running

9.1 Introduction

At present, coal seam and equipment are modeled separately [1–6], and the two are not organically combined, which leads to a large gap between the running state of equipment presented by virtual simulation and the actual situation, and also makes the research on virtual working face lose the real practical significance of industrial application. Therefore, it is necessary to display the fully mechanized mining equipment and coal seam in a digital way, and then compile the relevant behaviors to realize the simulation of the equipment in the actual coal seam. Although the relevant virtual reality simulation technology of fully-mechanized mining has made great breakthrough, it is not difficult to find that the operation simulation ability of equipment is poor based on coal seam modeled with 3DMine, ArcGIS and other mining engineering software [7, 8]. However, the equipment virtual simulation is mostly based on the virtual reality simulation software such as Unity3D and EON, and virtual coal seam cannot be created in the software. Therefore, there is no good method to achieve the modeling of equipment and coal seam as well as the integration of the two in software. The underground environment is bad. And the roof and floor environment is fluctuating. How to achieve the coupling between the three machine equipment and the roof and floor, the self-adaptive coordinated propulsion of the equipment under the control system and truly restore the underground equipment operation and coal seam morphology changes have become the research difficulty and hot spots.

It is necessary to improve the simulation credibility and simulation accuracy. Because of the above reasons, the following two technical problems must be solved:

- The preliminary simulation under the condition of ideal horizontal floor is relatively simple, but the actual simulation must consider the laying of scraper conveyor on the undulating floor, the real layout of hydraulic support group, the real walking of shearer on the scraper conveyor and the real recurrence and effective analysis of cutting curve;

X. Wang et al., *Virtual Reality Technology in Mining Machinery*,
https://doi.org/10.1007/978-981-16-4408-5_9

- Due to the lack of expression method of complete equipment and coal seam in virtual environment, the two models cannot be effectively linked. Some scholars have proposed to locate the equipment operation through advancing path and coordinate data. The main problems are the lager number of equipment and the fixed coal seam which cannot be updated. In addition, the computer real-time operational pressure caused by the problems is too high, which cannot effectively solve this problem. Therefore, it is urgent to develop a new method or software for virtual simulation fusion.

9.2 Overall Research Framework

In the virtual environment, how to establish a virtual production system for self-organized operation of equipment is an important goal of this study, and it can also provide an effective analysis tool for the study of space–time movement relationship of equipment operation.

The main problems to be solved include:

- Virtual equipment and virtual coal seam model establishment;
- Establish a stable and reliable interaction relationship between the virtual equipment and the coal seam model, and reflect it in the operating state of the equipment;
- High-precision simulation between each equipment;
- After the action of equipment and coal seam, the relationship between equipment and coal seam cutting, transportation and support can be updated in real time.

For this purpose, the research framework described in this paper is shown in Fig. 9.1. The functional relationship between equipment and coal seam is shown in Fig. 9.2.

The basic operation basis of this method is to determine the attitude of the scraper conveyor based on the cutting track of the shearer bottom plate, and further determine the position and pose information of the hydraulic support in the process of propulsion. The cutting track of the roof determines the support space of the hydraulic support and further determines the drop column quantity of the hydraulic support in the process of moving the support. From the perspective of time dimension, the results of cutting the top and bottom of the shearer have a direct impact on the subsequent actions of the scraper conveyor and hydraulic support, and also reflect the interaction between equipment and coal seam [9].

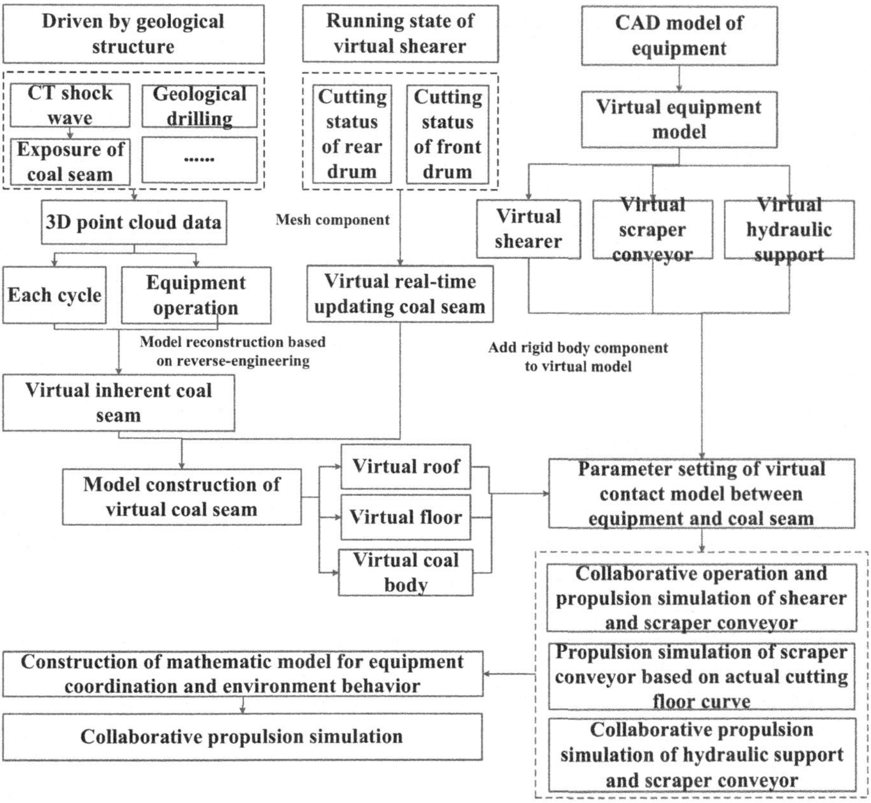

Fig. 9.1 Overall technical concept and framework

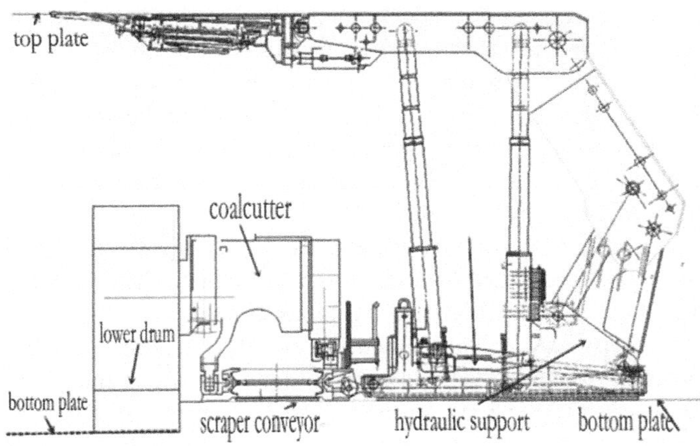

Fig. 9.2 Connection relationship of three machines in a fully mechanized coal mining face

9.3 Establishment of Virtual Equipment and Coal Seam Model

The establishment of virtual equipment model has been relatively mature [10], and the establishment of virtual coal seam model is mainly solved here. This paper proposes that the virtual coal seam mainly includes the virtual inherent coal seam and the virtual real-time updating coal seam. The established methods mainly include the reverse reconstruction method and the virtual software direct construction method, and the comprehensive analysis is carried out to complete the coal seam modeling combined with the characteristics of the two.

9.3.1 Two Basic Methods of Coal Seam Modeling

The first method by using geological data to construct three-dimensional point cloud, after format conversion enter Unity3D software, transform method is: the ArcGIS software → Txt three-dimensional point cloud → ImageWave(UG) software → STL format → 3DMAX software → fang format → Unity3D software. The second method is to directly construct coal seam in Unity3D, using Mesh grid and LineRender technology.

The reverse reconstruction method is mainly used to scan the shape and structure of the object and obtain the spatial coordinates of the object surface. It can transform the three-dimensional information of the object into a digital model that can be directly processed by the computer. It provides a very convenient and fast means for the digitization of the object.

LineRenderer is mainly used to render line segments in Unity3D software. Compared with traditional GL image library to render line segments, it has more powerful functions and is easier to set properties such as color and width. Mesh components are actually triangular meshes, which are polygonal meshes composed of a series of triangles. They are mainly used to simulate the surface of complex objects. In virtual reality systems, they are often used to simulate various contact objects and object navigation. Coupled with a physical simulation engine, collision and contact behavior can be simulated.

The comparison of the two modeling methods in modeling speed, operation speed, flexibility and collision characteristics is shown in Table 9.1. The model established by the first method is relatively fixed and the operating pressure of the computer is low. The model of the second method is more flexible, but it will increase the operating pressure of the computer in grid building. Therefore, the advantages and disadvantages of the two methods are comprehensively analyzed. The first method is used to build virtual inherent coal seam by block and cycle modeling, and the second method is used to build virtual real-time updating coal seam. The whole virtual coal seam model can achieve real-time correction by controlling whether the attributes are activated or not.

Table 9.1 Comparison of two coal seam modeling methods

Methods of coal seam modeling	The modeling speed	Computing speed and computer pressure	Flexibility and dynamic data drive ability	Collision characteristics setting
Reverse reconstruction method	Quick	Low	Bad	Fast
Direct establishment method	Relatively slow	Larger	Optimum	Faster

9.3.2 Establishment of Virtual Inherent Coal Seam

The terrain detection points are constructed into three-dimensional space point cloud and the virtual coal seam three-dimensional space point cloud by using the existing geological exploration data of fully mechanized mining. According to every knife and cycle, multiple virtual coal seam data grid files are constructed. Each file is connected and spliced into virtual coal seam, and then the virtual inherent coal seam is formed.

The position of each file in the intermediate conversion process is consistent, which forming a whole inherent coal seam. However, the attributes of each part can be controlled by command. For example, if the gap between the actual cutting path and the coal seam interface is too large, the attribute of the coal seam file can be controlled as not activated, then the part of coal seam will not play a role. As shown in Fig. 9.3.

The virtual inherent coal seam includes virtual roof, virtual floor and virtual coal body. The virtual roof and the virtual floor are added with Collision module of Mesh components. The virtual coal body is gradually hidden according to the cutting cycle of the shearer.

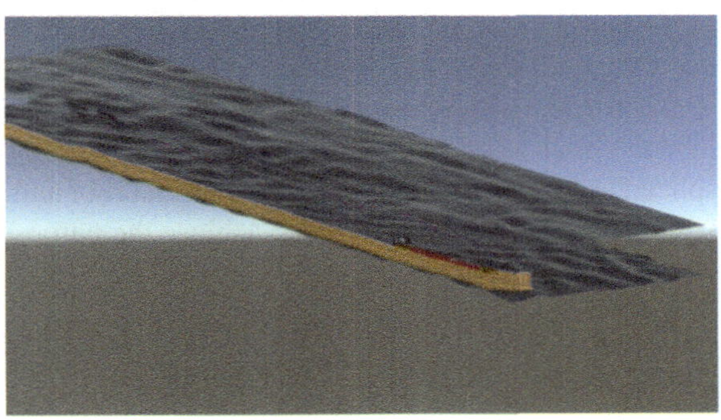

Fig. 9.3 Establishment of inherent coal seam

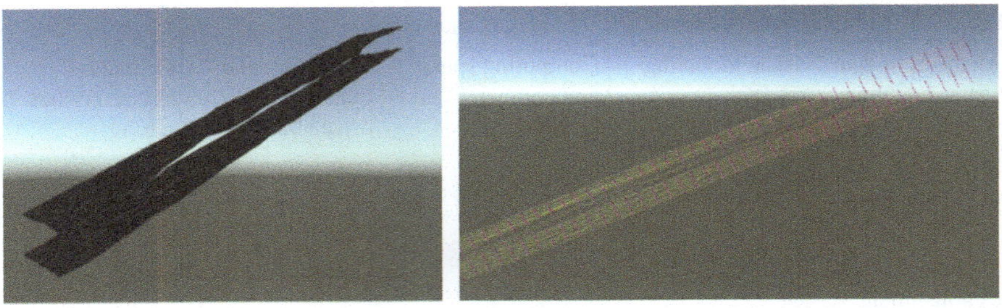

Fig. 9.4 Modeling method of virtual real-time updating coal seam

9.3.3 Establishment of Virtual Real-Time Updating Coal Seam

The construction of virtual coal seam with Mesh grid and LineRender technology generates XML format document based on real geological data, which reads data points in Unity3D and generates real coal seam surface by using Mesh grid. Then it uses LineRender component and Mesh component to connect with triangular grid to generate 3D roof and floor, and adds Mesh component collision body module in parallel. Coal seam can be driven to change by changing XML data points (Fig. 9.4).

9.3.4 Coal Seam Modeling Method with Bidirectional-Driving

The modeling of virtual coal seam should be combined with characteristics of the former two methods for bidirectional-driving. The inherent coal seam should be established in blocks before the simulation, and it should have certain flexibility. The updating method mainly uses the second method. Every time in a cycle, the curves of the cutting roof and the cutting floor of the front drum are recorded in real time, the XML coal seam data information is updated in real time, and the coal seam grid is constructed with real-time update. The attribute of the corresponding numbered inherent coal seam is controlled as not activated by command. Then the two are spliced and repaired to construct new virtual roof and floor in real time, and the coal seam is gradually hidden, so as to achieve the purpose of real-time update.

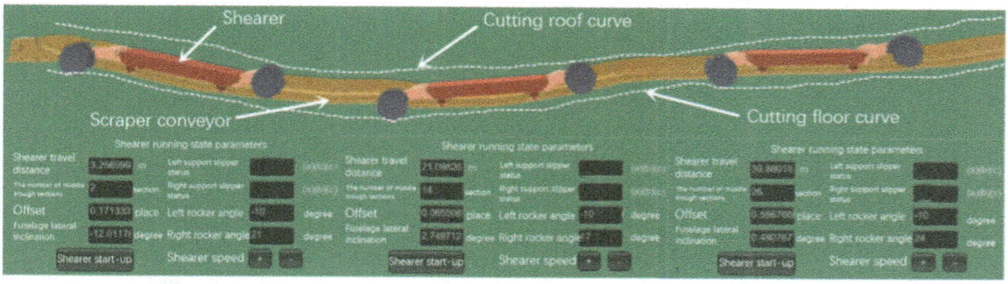

(a) (b) (c)

(a)-Downhill section (b)-Uphill section (c)-Horizontal section

Fig. 9.5 Operation state of Shearer in different condition

9.4 Operation Simulation Method Among Equipment

9.4.1 Cooperation Between Shearer and Scraper Conveyor

The virtual scraper conveyor interacts with the virtual coal seam to produce the suitable shape for the floor conditions. The shearer needs to ride on the scraper conveyor, so the virtual shearer must adapt to the shape of the virtual scraper conveyor. It is necessary to analyze the connection and constraint relationship between them. Through the real-time acquisition of the shape data of scraper conveyor and the coupling relationship between the left and right support slippers of shearer and the shovel plate of scraper conveyor and the coupling relationship between left and right walking wheels and pin row, the relationship between shearer fuselage position, pitch angle and scraper conveyor shape is solved, which provides reference conditions for shearer front and rear drum cutting virtual coal seam. The mathematical model is programmed into the bottom layer of virtual reality for real-time calculation, and the calculation results are driven to run the virtual model in real time. Figure 9.5 shows the virtual simulation of shearer running on scraper conveyor.

9.4.2 Scraper Conveyor and Hydraulic Support Coordination

The connection between the pushing mechanism of the hydraulic support and the middle trough is a floating system [11, 12], as shown in Fig. 9.6. The simulation between the virtual scraper conveyor and the virtual hydraulic support is mainly connected through the virtual hydraulic support pushing mechanism. Based on the analysis results of pushing mechanism structure, the background program is programmed to make the virtual scraper conveyor and virtual hydraulic support self-adaptive in different virtual coal seam conditions.

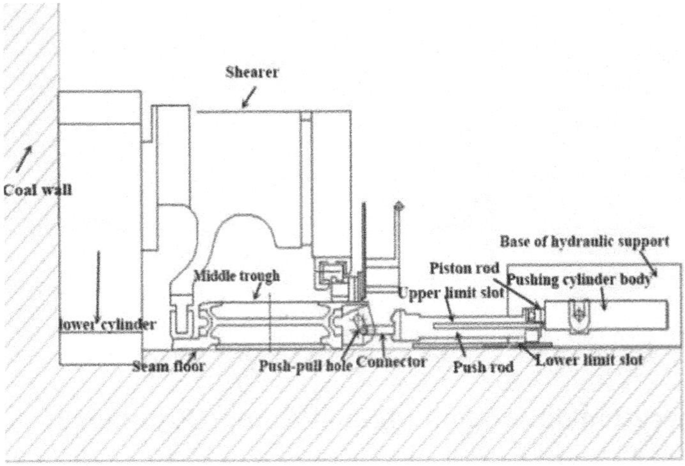

Fig. 9.6 Connection relationship between pushing mechanism of hydraulic support and scraper conveyor

The key position of the hydraulic support and scraper conveyor can be determined in the process of pushing, and advancing. The parameters of the pushing mechanism are determined, including the maximum bottom lifting amount, the maximum bottom sinking amount and the maximum offset. The coordinate changes of key points on the pushing mechanism are calculated. In order to deal with the change of floating connection under the condition of single or multiple hydraulic supports and different terrain conditions, the coordinate change of the connection position with the chute hole of the middle trough is mainly determined, and the extension amount of the pushing cylinder and the change of the angle of the pushing mechanism are determined. This paper studies the moving process under different cutting floor conditions, and uses the operation information determined by the action between the middle trough and coal seam and the operation information determined by the action of hydraulic support and coal seam, and analyzes the structure, so as to determine the connection operation state. The research results are programmed into the bottom layer of virtual reality software, which can complete the system of scraper conveyor and hydraulic support.

9.4.3 Collaboration of Shearer and Hydraulic Support Group

Different from the simulation of the ideal horizontal position of the floor, collaboration between shearer and hydraulic support structure must consider the relationship between the real-time position and attitude information of the shearer and the position of each hydraulic support, accord to the real-time cutting position of the front and rear drums to adjust process parameters, make the hydraulic support group coordinate with each other.

9.5 Establishment of Interaction Model Between Coal Seam and Equipment

9.5.1 Action Model of Coal Seam and Equipment

Using the built-in physical engine of Unity3D, and combining rigid body (Rigidbody), role Character (Character joint) and collision body (Collider) to complete the establishment of equipment and coal seam action model.

The specific steps are: perform physical simulation engine definition model patches → add components such as Mesh renderer, Rigidbody, Mesh Collider and Box Collider → select and set related parameters → read XML coal seam data by point and column to build virtual coal seam roof and floor → add Mesh Filter component to virtual coal seam →add the Mesh Filter component and Mesh Collider component to the virtual coal seam → complete the action between equipment and coal seam.

9.5.2 Hydraulic Support and Coal Seam Floor

The collaborative simulation process of virtual hydraulic support and virtual coal seam is as follows: add rigid bodies and collision components that are exactly the same shape as the base on the hydraulic support base model (Fig. 9.7) → analyze the key points of base and coal seam floor → interact with the virtual coal seam

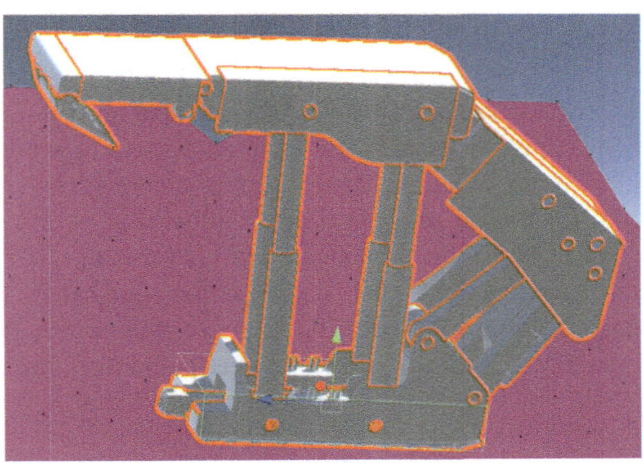

Fig. 9.7 Repairing model of a hydraulic support

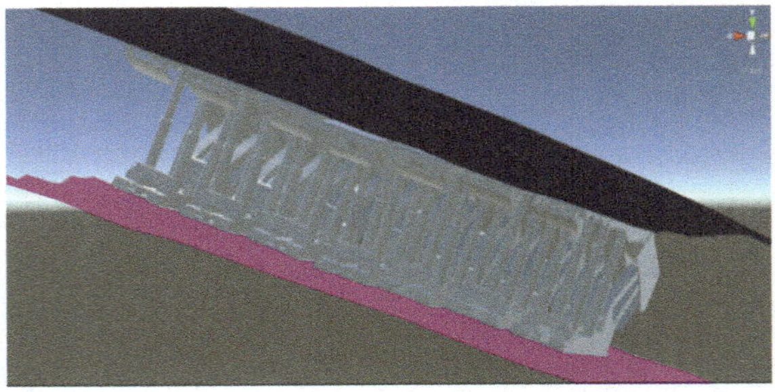

Fig. 9.8 Coupling of hydraulic support groups with roof and floor

floor → analyze the posture of the base and the bottom plate → after adapting to the bottom plate, analyze the posture of the coal seam roof and the roof of the hydraulic support → perform contact analysis on the roof of the virtual hydraulic support and the virtual coal seam roof → calculate the contact point and angle between the coal seam roof and the roof of the hydraulic support, according to the height of the top plate, calculate the rear link angle of the four-bar linkage → adaptive adjustment of the top and bottom plate when the hydraulic support is advanced → determine the gesture of the hydraulic support.

Then establish the simulation of the hydraulic support group. The simulation of the virtual support and the surrounding virtual support is to add a cube collider in the working space of each virtual support from the coal seam floor to the coal seam roof. In the process of running simulation, the interference between virtual hydraulic support and surrounding hydraulic support such as squeezing frame and biting frame immediately triggers alarm script and gives corresponding instructions. In the process of virtual advancing path, the hydraulic support must also have the ability to sense the running state of the surrounding hydraulic support, and cooperate to complete the virtual roof support task, which requires real-time adaptation and calculation, and gradually coupled with the roof curve. The simulation effect is shown in Fig. 9.8.

9.5.3 Scraper Conveyor and Coal Seam Floor

The simulation of virtual scraper conveyor and virtual coal seam is to repair a rigid body component which is completely consistent with the bottom shape at the bottom of each virtual middle trough, so that it has rigid body property and acts with virtual coal seam. The adjacent virtual middle troughs are connected by hinges, and the connection parameters are adjusted, so that the whole scraper conveyor can fully function with the virtual coal seam floor, and can be laid on the floor adaptively, so as to obtain the attitude data of the scraper conveyor on the floor, and provide the running track for the virtual shearer, as shown in Fig. 9.9.

Fig. 9.9 Self-adaptation of the scraper conveyor and the floor

Fig. 9.10 Cutting roof and floor curve of shearer

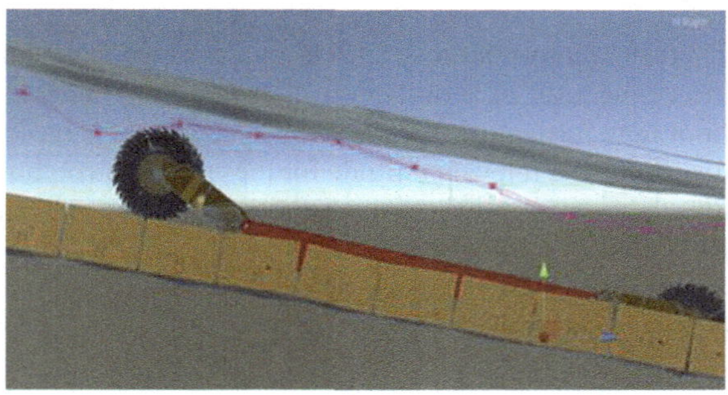

9.5.4 *Shearer Cutting Coal Seam*

The simulation of virtual shearer and virtual coal seam is that the virtual shearer cuts the virtual coal seam by reciprocating walking. Through the analysis of the virtual shearer's walking direction, drum height adjusting speed and position information, combined with the preset cutting path, the height adjustment and cutting path tracking are carried out, as shown in Fig. 9.10. The cutting points of roof and floor are recorded in real time and stored in XML. The cutting curve is generated by reading the points of bottom and top beds in real time by Mesh network and connected with the points of the previous cycle to construct Mesh network. Meanwhile, the Mesh network updates the virtual coal seam model in real time.

9.6 System Integration Technology Path

Integration is to experiment and practice the above technologies and methods, and integrate them after verifying these are feasible. It is not simply integrated together.

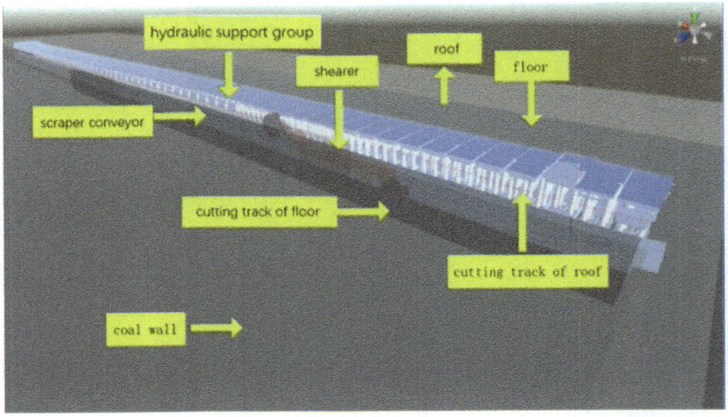

Fig. 9.11 Conception of joint virtual simulation integration of coal seam and equipment

There are many conflicts, such as the moving mechanism of hydraulic support group, the connection of scraper conveyor and the connection between virtual coal seam floor may form a insufficient DOF system. Some basic theories need to be studied to determine the relevant attitude for these above problems. The static characteristics of these above researches have been tested, and the dynamic operation characteristics need to be further tested.

After the above work, it is necessary to establish and improve the mathematical model of the coordination between fully mechanized coal winning equipment. After the virtual equipment is loaded with the corresponding script, it can independently simulate the state of the timely advance between the fully mechanized coal winning equipment and the coal seam. The current status of the research is shown in Fig. 9.11.

After the completion of the joint virtual simulation of equipment and coal seam, the next step will be to increase the research of the theory that equipment joint positioning and attitude monitoring, especially after accessing the real-time operation data of equipment, the bottom's model is still faced with problems such as ambiguity and uncertainty, and may produce redundant information, leading to the situation that the real state of equipment and coal seam cannot be accurately determined. Therefore, it is necessary to establish a series of conflict resolution schemes to achieve real-time correspondence between real-time data of high fidelity virtual working face and physical comprehensive mining working face, so as to build a real-time data-driven three-dimensional scene reproduction and remote intervention integrated operation platform.

9.7 Conclusions

This paper puts forward the concept of "coal seam + equipment" joint virtual simulation operation technology in comprehensive mining working face, and makes corresponding strategies and practices. Compared with the existing technology, it has the following advantages:

- This technology introduces virtual reality physical engine in the process of virtual simulation of fully mechanized coal mining, which can make virtual equipment act on virtual coal seam, and then simulate the operation information of actual equipment on coal seam. It can provide high-precision information of coal seam action, replace the method of coordinate propulsion, complete the high simulation promotion under the three-dimensional space virtual coal seam environment, and completely reproduce the whole comprehensive mining working face running process.
- This technology can update the cutting status of equipment and the change state of coal seam in real time, which can completely present the timely status and general situation of coal mining change in extraction process.
- It provides a new idea, strategy and method for studying the space–time kinematic relationship of comprehensive mining working face.

References

1. Mao, S.J., Yang, N.S., Gao, Y.Q., et al.: Design and key technology research of coal mine distributed cooperative "one map" system. J. China Coal Soc. **43**(1), 280–286 (2018)
2. Jia, Q.R., Chen, D.F., Li, J.X., et al.: Three-dimensional modeling method of coal seam with gradual refinement. J. Northeast. Univ. Nat. Sci. **39**(5), 726–730 (2018)
3. Wang, A.M.: Application of 3D visualization modeling technology in mine design. Saf. Coal Mines **49**(11), 129–132 (2018)
4. Liu, G.W., Cheng, G., Song, J.C., et al.: Study on integrated construction technology of noncontinuous coal seam solid model. Coal Sci. Technol. **46**(05), 210–216 (2018)
5. Zhou, W.X., Chen, Y.H., Yang, Y.G., et al.: 3D modeling and visualization of coal reservoir based on corner-point grid. Coal Geol. Expl. **44**(5), 53–57 (2016)
6. Xie, J.C., Yang, Z.J., Wang, X.W., et al.: Research on key technologies of virtual collaboration of three machines in fully mechanized coal mining face. Chin. J. Eng. Des. **1**, 85–93 (2018)
7. Ralston, J.C., Reid, D.C., Dunn, M.T., et al.: Longwall automation:Delivering enabling technology to achieve safer and more productive underground mining. Int J Min Sci Techno. **25**(6):865–876 (2015)
8. Li, W., Luo, C., Yang, H., et al.: Memory cutting of adjacent coal seams based on a hidden Markov model. Arab. J. Geosci. **7**(12), 5051–5060 (2014)
9. Ge, S.R., Wang, Z.B., Wang, S.B.: Study on key technology of internet plus intelligent coal shearer. Coal Sci. Tech. **44**(7), 1–9 (2016)

10. Xie, J.C., Yang, Z.J., Wang, X.W., et al.: Research of seamless linkage method for hydraulic support components under VR environment. Chin. J. Eng. Des. **24**(4), 373–379 (2017)
11. Xie, J.C., Yang, Z.J., Wang, X.W., et al.: A joint positioning and attitude solving method for shearer and scraper conveyor under complex conditions. Math. Probl. Eng. **2017**, 3793412–3793426 (2017)
12. Niu, J.F.: Research of straightness control system of fully-mechanized coal mining face. Ind. Mine Autom. **41**(5), 5–8 (2015)

Chapter 10
A Virtual Reality Collaborative Planning Simulator Based on a Multi-agent System

10.1 Introduction

In Industrial 4.0, SIEMENS first proposed the so-called digital win model, in which all elements in the actual production process are digitalized, and all involved production processes can be simulated and analyzed within this digital level. This model can predict the events that are likely to occur if the actual production runs and help avoid unnecessary investment losses before production begins. However, the application of VR technology remains relatively rare in coal-mining equipment, and a large gap remains to be filled in genuinely promoting the development of coalmining equipment.

A fully mechanized working face is the most critical link in the coal industry. Therefore, based on the above analysis, this paper will establish a VR simulator that can solve the problem of the fully mechanized coalmining face and carry out visual planning details. This advancement will help with decision making in the conceptual design, design selection and actual operation stages of the mining project. After analyzing the application and development of fully mechanized coalmining equipment and VR technology in detail, a fully mechanized Unity3D simulator (FMUnitySim), which can plan and adjust the three machines' key parameters online, is proposed in terms of mathematics and artificial intelligence.

10.2 Framework of FMUnitySim

10.2.1 Overall Framework

Based on the established collaborative mathematical model of three machines in a dynamic environment and the MAS theory, FMUnitySim (Fig. 10.1) defines the virtual behaviors of three machines and their interactions with the virtual environment

X. Wang et al., *Virtual Reality Technology in Mining Machinery*,
https://doi.org/10.1007/978-981-16-4408-5_10

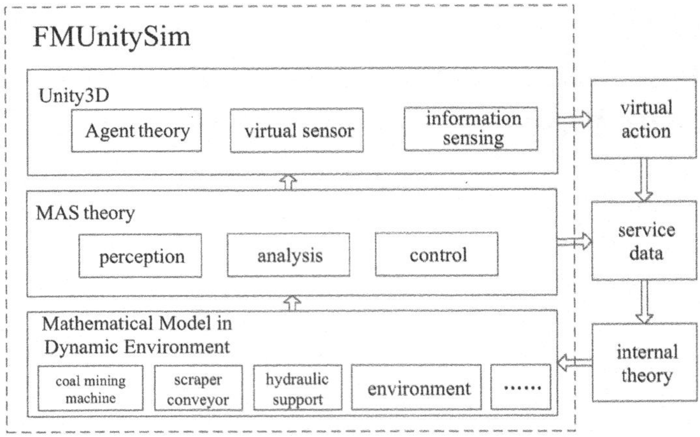

Fig. 10.1 Overall framework of VR collaborative planning environment

using C# script in Unity3D, thereby enabling the visualization of the entire planning process and enabling all relevant process data to be acquired and analyzed.

10.2.2 Collaborative Mathematical Model of Three Machines

The entire coordination process of the three machines is in accordance with the shearer location and relevant regulations. The hydraulic supports in different areas will take diverse but coordinated actions: those in the front of the shearer flap out, whereas those at the rear of the shearer slide in front of the support, which constitutes the support of the suspended roof and coal wall after coal cutting. After the sliding advance of the support, the hydraulic supports will push the scraper conveyor close to the coal wall side and load the falling coal to be shipped out. The overall goal of coordinating these three machines is as follows:

- Automatically move the coalmining equipment.
- Avoid any interference between the shearer and hydraulic support.
- Maintain and guarantee the suitable running posture and straightness of the scraper conveyor.
- Effectively manage the coal wall and roof of the coalmining face.
- Ensure that the hydraulic support strength attain the initially specified value.

To achieve these goals, the three machines must work in close coordination and achieve effective interaction with the underground environment. Figure 10.2 shows the influencing factors in the three machines for different geological conditions at the coalmining face.

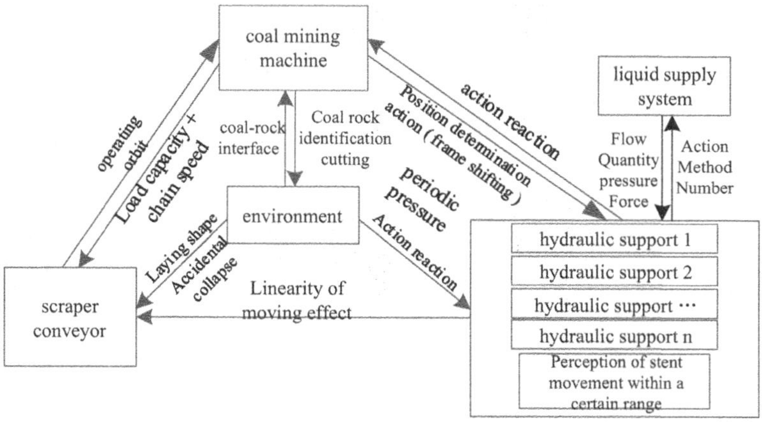

Fig. 10.2 Analysis on interaction factors of cooperative operation

- The coordinated operation of the three machines requires the ability of comprehensive organization, with which the information from the ever-changing underground environment, mutative equipment conditions, supplier devices, running speed of shearer, and scraper conveyor load can be accordingly synthesized, optimized, and controlled.
- The coordinated operation of three machines requires adaptation and optimization capabilities with which the shearer haulage speed and numbers and parameters of the advancing mode can be automatically adjusted according to various production scenarios and roof conditions.

10.2.3 Collaborative Planning Model Based on an MAS

As the core of multi-agent analysis, coordination and cooperation enable the knowledge, expectations, intentions, planning, and actions to collaborate with one another. In other words, the coordination in an MAS is a process in which all agents interact with one another to achieve a common goal in a compatible and harmonic manner and avoid deadlocks or mutual locks among the agents.

Therefore, in this study, the shearer, scraper conveyor, grouped hydraulic supports, hydraulic system and underground environment are taken as an agent; they swap and sense information with one another, which affects and controls their behaviors. The interactions among all agents of the three machines and the environment used to effectively cut and transport coal are shown in Fig. 10.3.

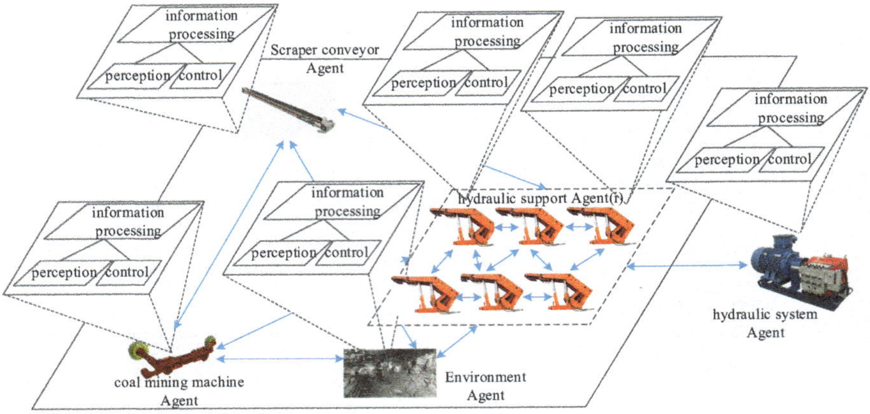

Fig. 10.3 Interaction diagram between 'three machines' Agent and underground environment

10.2.4 VR Planning Method

In the VR environment of Unity3D, virtual equipment can be controlled by a programmed C# script based on the collaborative mathematical model and collaborative planning model.

- With the control of CmjAgent.cs, the virtual shearer can realize coordinated motion between the rocker arm and vertical steering cylinders, control the shearer haulage speed and direction, and simulate the movement of the real shearer.
- With the control of GbjAgent.cs, the virtual scraper conveyor can be adaptively laid on the virtual floor with the ability to push itself to the coal wall side and virtually transport and detect the ship capacity of coal.
- With the control of YyzjAgent.cs, the hydraulic support can perform movement, such as setting the legs, retracting the legs, advancing the support, and pushing the conveyor.
- Under the control of YyxtAgent.cs, the hydraulic system can provide hydraulic oil for hydraulic supports according to different conditions.

With the control of EnvAgent.cs, the virtual environment can construct the virtual roof and floor from XML data, which are assigned by the users, to stimulate the mine pressure and broken roof.

10.3 Collaborative Mathematical Model of the Three Machines

10.3.1 Coupling of the Shearer Haulage Speed and Scraper Conveyor Load

With differences in the shearer haulage speed and location, the scraper conveyor typically changes its chain speed to maintain its rated load. Therefore, an accurate coupling analysis must be conducted, and the shearer haulage speed must be calculated for the scraper conveyor load.

10.3.1.1 Instantaneous Withstanding Maximum Load

The following research results can be obtained from Reference [1], and the maximum permitted coal transporting amount is calculated as:

$$Q_{permit} = \left(0.83 * N_{motor} - \frac{2q_0 L v_g \cos\beta f_1}{102\eta}\right)\frac{102\eta}{v_g(\cos\beta \pm \sin\beta)} \tag{10.1}$$

For the SGZ768/630-type scraper conveyor, the sectional area to transport coal is calculated according to Reference [1] as Area = 0.48 m^2.

The amount of coal stream per meter in the middle trough is calculated as:

$$A(t) = \begin{cases} \frac{Q(t)}{3.6K_g v_0} & \frac{Q(t)}{3.6K_g v_0} < A_{area} \\ A_{area} & \frac{Q(t)}{3.6K_g v_0} \geq A_{area} \end{cases} \tag{10.2}$$

10.3.1.2 Analysis of Cutting with the Coal-Transporting Process

For a shearer that cuts coal from its head to its tail, if the roof is assumed to be undulating and the floor is assumed to be flat, then the transporting coal process is as shown in Fig. 10.4.

The conversion conditions of four phases must satisfy the conditions shown in Table 10.1.

10.3.1.3 Calculation of the Coal Cutting Amount

We assume that a good coal-rock recognition device is installed in the shearer or that the operators can clearly distinguish the interface between coal and rock; then, the cutting amount of the front drum can be approximated as the full diameter of the

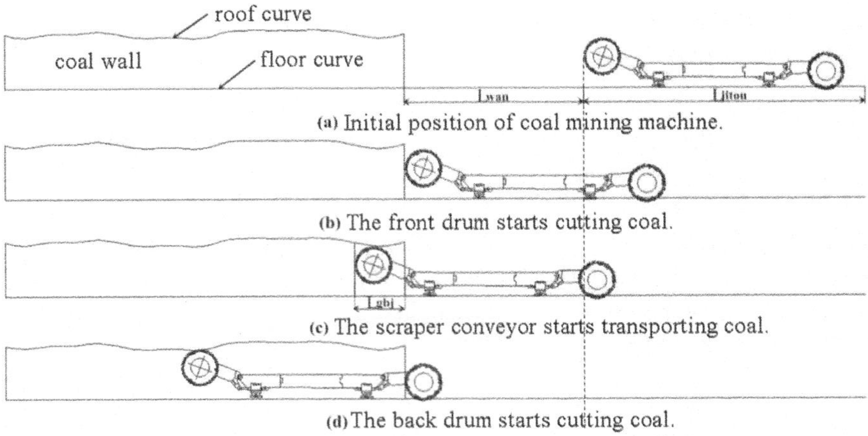

Fig. 10.4 Analysis of coal transportation process

Table 10.1 Conversion conditions of four stages

Conversion stage	Condition
(a)→(b)	$S_f(t_1) - S_f(t_0) + (L_y \cos\alpha_{S(t_1)} - L_y \cos\alpha_{S(t_0)}) > L_{wan}$
(b)→(c)	$\begin{cases} S_f(t_2) - S_f(t_1) + (L_y \cos\alpha_{S(t_2)} - L_y \cos\alpha_{S(t_1)}) = L_{gbj} \\ \int_{t_2}^{t_1} V_g dt = L_{wan} + L_{jitou} \end{cases}$
(c)→(d)	$S_r(t_3) - S_r(t_0) > L_{wan} + L_{JiShen} + L_y \cos\alpha_{S(t_0)} + D_{drum}/2$

front drum, and the cutting amount of the rear drum can be calculated according to the cutting trajectory of the front drum, as shown in Fig. 10.5.

As shown in Fig. 10.5, the definition of the coal seam curve is a series of feature points of cutting height per length of the middle trough in the X coordinates. Thus, the feature points of the cutting height of the rear drum are (X(i), Hu(i)-Ddrum).

$$\text{Define}\ \lambda = (\text{Sr(t)} - \text{Sr(t3)})/\text{Dzbc and the remainder} = (\text{Sr(t)} - \text{Sr(t3)})\%\text{Dzbc}.$$

Because the coal seam height changes slowly, the cutting line that connects two adjacent cutting points can be considered a straight line. Accordingly, the cutting area of two adjacent cutting points can be considered a trapezoid.

The coal cutting amount of the front drum and rear drum can be calculated as shown in Eqs. (10.3) and (10.4), respectively.

$$m_{front} = (S_f(t) - S_f(t_1) + L_y \cos\alpha_{t_1} - L_y \cos\alpha_t) * D_{drum} * J * \rho_{\text{soild}} \quad (10.3)$$

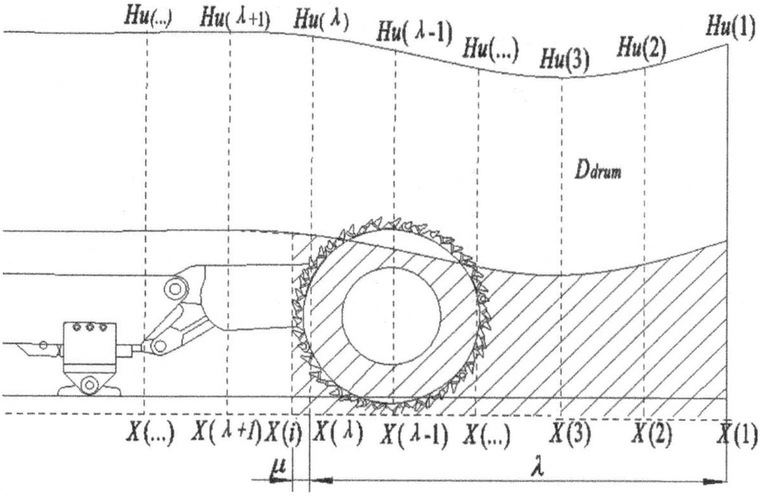

Fig. 10.5 Cutting curve of lower drum

$$m_{rear} = \left(\sum_{i=1}^{\lambda-1} (h_u(i) + h_u(i+1) - 2 * D_{drum}) * D_{zbc}/2 + \left(\frac{(h_u(\lambda+1) - h_u(\lambda))\sigma}{D_{zbc}} + 2 * (h_u(\lambda) - D_{drum})\right) * \sigma/2\right) * J * \rho_{soild} \tag{10.4}$$

The instantaneous coal cutting amount of the front and rear drums is calculated as shown in Eqs. (10.5) and (10.6) respectively.

$$m_{Ins\text{-}front}(t) = V_c * D_{drum} * J * \rho_{\text{soild}} \tag{10.5}$$

$$m_{Ins\text{-}rear}(t) = V_c * f(s_r(t)) * J * \rho_{\text{soild}} \tag{10.6}$$

The instantaneous amount of shipped coal is calculated as

$$m_{Ins\text{-}transport}(t) = q(t) * v_g * \rho_{dispersion} \tag{10.7}$$

The total amount of shipped coal from moment t2 to moment t is calculated as

$$m_{total\text{-}transport} = \sum_{i=t_2}^{t} m_{Ins\text{-}transport}(i) \tag{10.8}$$

The instantaneous load of the scraper conveyor is calculated as

$$Q(t) = m_{front}(t) + m_{rear}(t) - m_{total\text{-}transport}(t) \leq Q_{permit} \tag{10.9}$$

If $Q(t)$ reaches the maximum allowed value Q_{permit}, the scraper conveyor load at moment t must satisfy the following condition:

$$Q_{Ins}(t) = m_{Ins\text{-}front}(t) + m_{Ins\text{-}rear}(t) - m_{Ins\text{-}transport}(t) < 0 \tag{10.10}$$

10.3.1.4 Coupling of the Scraper Conveyor Load with the Underground Environment

In the hydraulic support process, the coal wall may suddenly collapse depending on the shearer location and running speed.

Whether collapse occurs mainly depends on the follower distance and mining height, and the common regulation is as follows: a higher mining height and greater follower distance correspond to a higher probability of coal wall collapse.

If this phenomenon occurs, the scraper conveyor load will change suddenly.

After this phenomenon occurs, it will not occur again within 50 m.

In this paper, assuming a scope of 50 m, if the follower distance is greater than 10 times D_{zbc}, the probability of collapse is calculated as follows:

$$f(p) = \begin{cases} 1 \ D_{follow} \geq 10 * D_{zbc} \\ 0 \ D_{follow} < 10 * D_{zbc} \end{cases} \tag{10.11}$$

where D_{follow} is the follower distance. The mutation load of the scraper conveyor is calculated as

$$\text{msudd} = (-1)\text{f(p)} * \text{Random.Range}(0.75, 1) * 2.5\text{t} \tag{10.12}$$

Considering the sudden collapse of the coal wall, the instantaneous load of the scraper conveyor is calculated as

$$Q(t) = m_{front}(t) + m_{rear}(t) - m_{total\text{-}transport}(t) + m_{sudd} \leq Q_{permit} \tag{10.13}$$

10.3.2 Coupling of the Shearer Haulage Speed and Adjustment of the Front Drum Height with the Underground Environment

Facing an undulating virtual roof, the virtual shearer can adjust the height of the front drum and adaptively plan according to its capacity.

Whenever the shearer runs along the length of the middle trough, the shearer will acquire the next roof height in advance and make a comparison with the current drum height. The shearer decides which action to take next according to the comparison result and ability of the hydraulic system.

When the shearer is operating at moment t, the rotate angle of the front drum is calculated as

$$\alpha_{S(t)} = \arcsin\left(\frac{H_u(i) - H_c - D_{drum}/2}{L_y}\right) \tag{10.14}$$

Due to the control strategy of the shearer or detected error of the coal seam height, the front drum is likely higher than the interface between coal and rock, and the front drum begins to cut the rock. In an abnormal electric current condition, the virtual shearer must reduce the height of the front drum.

The decision condition to determine whether the current is abnormal is as follows:

$$L_y \sin\alpha_i + H_c + D_{drum}/2 > H_{ui} \tag{10.15}$$

10.3.3 Coupling of the Following Control of Hydraulic Supports and the Shearer Haulage Speed

10.3.3.1 Matching the Advancing Mode

The most important factor of the three-machine coordination is the matching of shearer haulage speed V_c and hydraulic support advancing speed V_y.

If $V_c < V_y$, the hydraulic support can follow the shearer action in a sequential advancing mode, where the phenomena of support loss and inadequate advancing do not appear. Before the shearer location triggers the next group action, the current group has completed advancing the action.

If $2V_y > V_c > V_y$, the hydraulic support cannot regularly follow the shearer action in a sequential advancing mode and must switch to the cross-grouping advancing mode under the premise of a favorable roof condition. The hydraulic supports can continue running in order only by the cross-grouping advancing mode.

In the course of shearer running, the swarm hydraulic supports automatically switch the advancing mode according to the detected shearer haulage speed and follower distance in real time.

YiJiaFangShi, which is a static variable defined in FMUnitySim, is used to mark the current advancing mode. N, which is a static variable defined in FMUnitySim, is used to mark the number of hydraulic supports, which conduct advancing action. If the value of YiJiaFangShi is equal to one, the current advancing mode is the sequential advancing mode, and the value of N is unique. If the value of YiJiaFangShi is equal

Table 10.2 State mark table of hydraulic support

Stand state	Pick up the guard panel	Fall pillar	Mobile rack	Rising pillar	Roof support	Push the slip
State	1	2	3	4	5	6

to two, the current advancing mode is the cross-grouping advancing mode, and N is a smaller number.

10.3.3.2 State Mark of the Hydraulic Support Using the Finite-State Machine

Hydraulic support has six states, which are denoted as 1–6, as shown in Table 10.2.

The hydraulic support must detect the distance between its position and the position of the front drum or rear drum. If the distance satisfies the rules shown below, the hydraulic support will take the corresponding actions.

- Rule 1: if the distance from its position to the position of the rear drum is less than two times Dzbc, the hydraulic support begins to successively retract the columns, advance the support and rise the columns.
- Rule 2: if the distance from its position to the position of the rear drum is within 10–15 m, the hydraulic support begins pushing the conveyor.
- Rule 3: If the distance from its position to the position of the front drum is less than two times Dzbc, the hydraulic support begins to flap out.

10.3.3.3 Realization of the Sequential Advancing Mode

If the value of YiJiaFangShi is equal to one, the current advancing mode is the sequential advancing mode. The conditional relation is as follows:

$$state(z) = \begin{cases} 1 & S_{zj}(z) - S_f(t) > 2 * D_{zbc} \\ 2 & (S_r(t) - S_{zj}(z) > (2-6) * D_{zbc}) \cap (state(z-1) = 5) \\ 3 & (state(z) = 2) \cap (H_u(z) - H_{zj}(z) \geq H_{down}) \\ 4 & (state(z) = 3) \cap (S_{tuiyi}(z) = J) \\ 5 & (state(z) = 4) \cap (H_u(z) - H_{zj}(z) \leq 0) \\ 6 & (S_r(t) - S_{zj}(z) > (7-10) * D_{zbc}) \cap (state(z) = 5) \end{cases} \quad (10.16)$$

10.3.3.4 Realization of the Cross-Grouping Advancing Mode

If the value of YiJiaFangShi is equal to two, the current advancing mode is the cross-grouping advancing mode. Suppose that hydraulic supports *No. z* and *No. z*

+ 2 maintain the cross-grouping advancing mode; then, hydraulic support No. z is calculated as shown in Eq. (10.16), and hydraulic support *No.* $z + 2$ must be synchronized with hydraulic support No. z. The conditional relation is shown below.

$$state(z+2) = \begin{cases} 1 & S_{zj}(z) - S_f(t) > 2 * D_{zbc} \\ 2 & state(z) = 2 \\ 3 & (state(z) = 3) \cap (H_u(z+2) - H_{zj}(z+2) \geq H_{down}) \\ 4 & (state(z) = 4) \cap (S_{tuiyi}(z+2) = J) \\ 5 & (state(z) = 5) \cap (H_u(z+2) - H_{zj}(z+2) \leq 0) \\ 6 & (S_r(i) - S_{zj}(z) > (7-10) * D_{zbc}) \cap (state(z+2) = 5) \end{cases} \tag{10.17}$$

10.3.3.5 Switching of the Advancing Mode

If the follower distance is greater than 11 times D_{zbc}, the shearer must decelerate. While waiting for the previous hydraulic support to finish advancing, the advancing mode switches from the sequential advancing mode to the cross-grouping advancing mode.

If the follower distance is less than 3 times D_{zbc}, the shearer must accelerate. Similarly, under the premise of waiting for the previous hydraulic support to finish advancing, the advancing mode switches from the cross-grouping advancing mode to the sequential advancing mode.

10.3.4 Coupling of the Following Control of Hydraulic Supports with the Condition of the Roof and Floor

When the underground environment is in a good condition, the advancing action of hydraulic supports must match the difference of the shearer haulage speed and location, which follow rules 1–3 in Sect. 10.3.2. For a working face with a broken roof, the advancing mode of the support with pressure is recommended to prevent the roof from falling. The column retracting speed for the working face with a large mine pressure is slower than that under normal conditions because the relieving the pressure requires a long period of time.

With the continuous operation of three machines, the action speed of the swarm hydraulic cylinders gradually decreases because of the filter plug or other problems, which makes the original control parameters unsuitable. These problems may occur at any time in any hydraulic support; thus, the action quality cannot be effectively controlled.

The action time of the advancing supports can be calculated as:

Table 10.3 Value table of working condition coefficient

Coefficient	Taking values	Conditions
n_{broken}	1.5–2.0	$B_{zj}(i) \geq B_{normal}$
	1	$B_{zj}(i) < B_{normal}$
n_{press}	1.3–1.5	$P_{zj}(i) \geq P_{normal}$
	1	$P_{zj}(i) < P_{normal}$
n_{hy}	1–1.1	$YiJiaFangShi = 1$
	1.3–1.5	$YiJiaFangShi = 2$
$n_{condition}$	1–1.15	The longer the running time, the larger the value

$$t_{norm-move} = (n_{broken}) * (n_{press})\left(n_{hy} * n_{condition} * \left(\frac{H_{rise}}{S_{r-l}} + \frac{H_{down}}{S_{d-l}} + \frac{J}{S_{r-t}}\right)\right) \quad (10.18)$$

The values of the operating parameters are shown in Table 10.3.

10.3.5 Coupling of the Shape of the Scraper Conveyor and Advancing Units of the Hydraulic Supports

When the shearer runs in the feeding process, the scraper convey will form an s-shaped bending section. When the shearer runs in the normal cutting process, the straightness of every middle trough must be detected in real time.

We assume that the roof is flat and that each hydraulic support maintains the normal attitude, i.e., it does not appear askew, and no inclination occurs in the support advancing process. Therefore, the s-shaped and straightness of the scraper conveyor depend on the elongation of the advancing units for the corresponding hydraulic supports.

10.3.6 Delay or Data Loss

In the process of interacting with other machines, each machine may encounter delay or data loss. The interaction of the three machines is as follows: the shearer occupies the lead position in the three-machine movement, and the scraper conveyor and group hydraulic supports are auxiliary.

The shearer may obtain the number of hydraulic supports, which delays advancing support and results in a greater distance at the empty top. The shearer may also obtain the scraper conveyor's single load overrun, which causes a short-time overrun of the scraper conveyor. Conversely, this effect decreases the efficiency of the coal mining process.

By obtaining the shearer's location in real time, the advancing action of a single hydraulic support is continuous. However, the hydraulic supports in front of the shearer may retract the supporting plate farther backward than is normal, which causes a risk of interference.

Those behind the shearer will struggle more than normal while advancing support. This widens the empty distance and increases the collapse risk of the roof coal wall.

10.4 Establishment of Agent

10.4.1 Model of the Shearer Agent

An agent model of the shearer is shown in Fig. 10.6. The perception and communication module is used to interact with other agents (Table 10.4); then, the logical

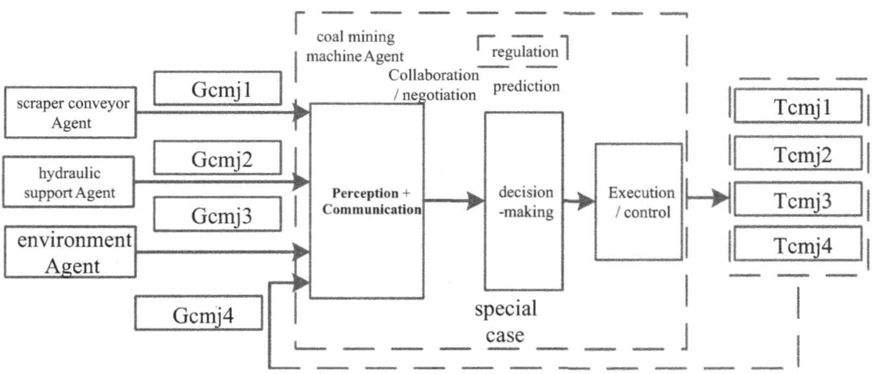

Fig. 10.6 Coal mining machine agent model

Table 10.4 Sensing task of coal mining machine and sensing variable of virtual coal mining machine

Serial number	Sensing task of coal mining machine	Sensing variable of virtual coal mining machine
1	Gcmj1: Coal rock identification task to obtain the coordinates of the next roof point in front of the drum	$H_u(i)$
2	Gcmj2: sensing scraper conveyor (overload, light load)	$Q(t)$, $Q_{Ins}(t)$
3	Gcmj3: sensing hydraulic support; (follow up, fall angle behind, surrounding rock condition)	N, D_{follow}
4	Gcmj4: perception of oneself	I_{motor}

Table 10.5 Control task of coal mining machine and control variable of virtual coal mining machine

Serial number	Control task of coal mining machine	Control variable of virtual coal mining machine
1	Tcmj1: expansion control of left cylinder (elongation, shortening, maintenance)	$\alpha_{S(t)}$
2	Tcmj2: expansion control of right cylinder (elongation, shortening, maintenance)	$\alpha_{r\,S(t)}$
3	Tcmj3: traction rate control (increase, decrease, maintain)	V_c
4	Tcmj4: update motion track control (railway recognition, transformation direction)	$S_{tuiyi}(1), \ldots, S_{tuiyi}(m), \ldots, S_{tuiyi}(m_{\max})$
5	Tcmj5: virtual current (whether overrun)	I_{motor}, $\alpha_{S(t)}$

information processing module is used to reason, decide and transfer results to the execution and control module (Table 10.5), which manipulates the virtual shearer to take specific actions.

The shearer haulage speed should be coupled with the conditions of the roof, scraper conveyor and hydraulic supports.

Coupled with the hydraulic supports, the virtual shearer must take corresponding actions when is facing the minimum and maximum distance of empty roof.

Coupled with the scraper conveyor load, if the scraper conveyor load is greater than the permitted maximum load, the instantaneous cutting amount of coal is not greater than the instantaneous transport amount of shipped coal.

Coupled with the roof plate, if the coal-rock interface identification is regularly working, a shearer should be in the normal condition. In contrast, when abnormal turning of the virtual electric current of motors occurs, the shearer must decelerate and decrease the height of the front drum by adjusting the front lifting cylinder.

10.4.2 Model of the Scraper Conveyor Agent

The corresponding relationship between the perception task of the scraper conveyor and the perception variables of the virtual scraper conveyor is shown in Table 10.6, and the corresponding relationship between the control task of the scraper conveyor and the virtual control variable of the scraper conveyor (GbjAgent.cs) is shown in Table 10.7.

If the scraper conveyor load is greater than the maximum permitting load, the instantaneous amount of coal cut by the front and rear drums and the instantaneous

Table 10.6 The corresponding table of scraper conveyor perception task and virtual scraper conveyor perception

Serial number	Scraper conveyor perception task	Virtual scraper conveyor perception task
1	Ggbj1: control of coal load detection device	$Q(t)$, Q_{Ins}
2	Ggbj(i)2: perception of left and right adjacent chute posture	$S_{tuiyi}(i-1)$, $S_{tuiyi}(i)$, $S_{tuiyi}(i+1)$

Table 10.7 The corresponding table of scraper conveyor control task and virtual scraper conveyor control

Serial number	Scraper conveyor control task	Virtual scraper conveyor control task
1	Tgbj1: chain speed control of scraper conveyor	V_g

amount of shipped coal must be detected online. If the former is greater than the latter, the shearer haulage speed must decelerate.

10.4.3 Model of the Hydraulic Support Agent

There are currently over 100 hydraulic supports in the actual underground working face. Each hydraulic support must communicate with other agents and have the appropriate behavior for the current underground environment, as shown in Fig. 10.7.

The corresponding relationship between the perception task of the hydraulic support and the perception variables of the virtual hydraulic support is shown in Table 10.8, and the corresponding relationship between the control task of the hydraulic

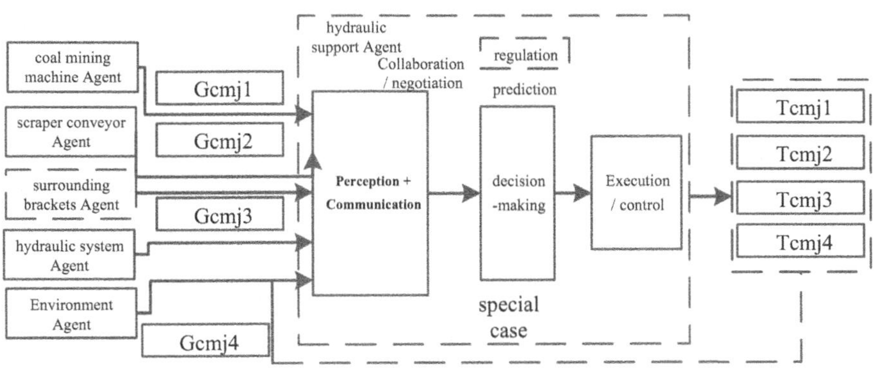

Fig. 10.7 Hydraulic support (i) Agent model

Table 10.8 The corresponding table of hydraulic support sensing task and virtual hydraulic support sensing

Serial number	Hydraulic support sensing task	Virtual hydraulic support sensing task	Qualitative or quantitative
1	Gzj(i)1: sensing coal mining machine control; (position and velocity)	$S_f(i)$, $S_r(i)$,V_c	Quantitative
2	Gzj(i)2: sensing m and the control of central trough within a certain range	$S_{tuiyi}(i)$	Quantitative
3	Gzj(i)3: sensing the hydraulic support control within a certain range adjacent to the left and right	*YiJiaFangShi*	Quantitative
4	Gzj(i)4: perceived emulsion pump flow pressure stability requirements control	n_{hy}, *YiJiaFangShi*	Quantitative
5	Gzj(i)5: sensing downhole environmental pressure condition control	$B_{zj}(i)$,$P_{zj}(i)$, n_{broken}, $n_{pressure}$	Quantitative
6	Gzj(i)6: perceiving self motion	$n_{condition}$	Quantitative

support and the control variable of the virtual hydraulic support (YyzjAgent.cs) is shown in Table 10.9.

Table 10.9 The corresponding table of hydraulic support control task and virtual hydraulic support control

Serial number	Hydraulic support control task	Virtual hydraulic support control task	Qualitative or quantitative
1	Tzj(i)1: Pick up the guard panel task; State = 1	$\phi(i)$ the angle of mutual support plate of No. i bracket	Qualitative
2	Tzj(i)2: Fall pillar task; State = 2	$H_{zj}(i)$	Quantitative
3	Tzj(i)3: mobile rack task; State = 3	$S_{tuiyi}(i)$	Quantitative
4	Tzj(i)4: rising pillar task; State = 4	$H_{zj}(i)$	Quantitative
5	Tzj(i)5: roof support task; State = 5	$\phi(i)$	Qualitative
6	Tzj(i)6: push the slip task; State = 0	$S_{tuiyi}(i)$	Quantitative

Table 10.10 The corresponding table of emulsion pump station (hydraulic system) perception task and virtual hydraulic system perception

Serial number	Emulsion pump station (hydraulic system) perception task	Virtual hydraulic system perception task
1	Grb1: perceptual hydraulic support action mode, action number, etc	*YiJiaFangShi*

Table 10.11 The corresponding table of emulsion pump station control task and virtual hydraulic system control

Serial number	Emulsion pump station (hydraulic system) control task	Virtual hydraulic system control task
1	Trb1: flow control task; (alarm)	n_{hy}
2	Trb2: pressure control task; (pressure loss)	n_{hy}

Hydraulic support (i) must detect the distance between its position and the front drum or rear drum. If the distance satisfies rules 1, 2 and 3, hydraulic support (i) performs the corresponding actions.

In the support advancing process, it is necessary to couple the roof trajectory, mine pressure, hydraulic system and surrounding hydraulic supports. The action of pushing the conveyor and advancing the support must be coupled with the shape of the scraper conveyor.

10.4.4 Model of the Hydraulic System Agent

The hydraulic system is mainly affected by the action mode and amount of swarm hydraulic supports. The corresponding relationship between the perception task of the hydraulic system and the perception variables of the virtual hydraulic system is shown in Table 10.10, and the corresponding relationship between the control task of the hydraulic system and the virtual control variable of the hydraulic system (YyxtAgent.cs) is shown in Table 10.11.

10.4.5 Model of the Underground Environment Agent

As shown in Fig. 10.8, the agent of the underground environment can simulate the change in the underground environment parameters because of the operation of the three machines and regular events in the underground environment, which include the collapse of the coal wall and broken roof caused by the empty roof distance and periodic mine pressure in the geology condition.

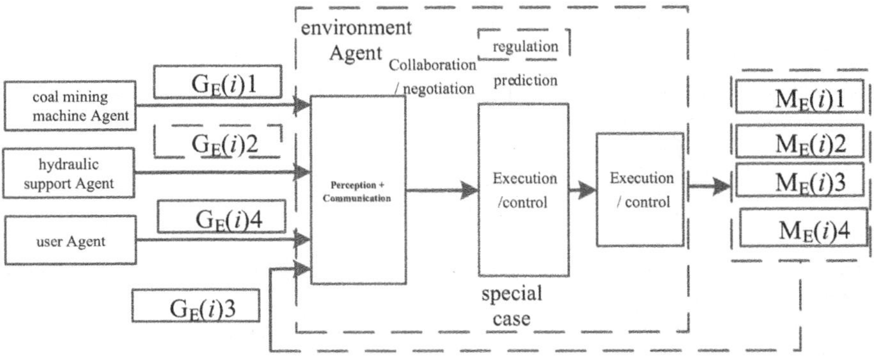

Fig. 10.8 Underground environment agent model

Table 10.12 The corresponding table of virtual underground environment perception task and virtual underground environment perception

Serial number	Virtual underground environment perception task	Virtual underground environment perception task
1	$G_E(i)1$: perceptual tracking distance	$S_r(i) - S_{zj}(N)$
2	$G_E(i)2$: sensing effect of hydraulic support on roof	$H_u(i) - H_{zj}(i)$
3	$G_E(i)3$: perceived periodic weighting, geological changes	$B_{zj}(i)$, $P_{zj}(i)$, n_{broken}, $n_{pressure}$
4	$G_E(i)4$: Underground environmental conditions of initial user input	$H_u(i)$

The corresponding relationship between the perception task of the environment and the perception variables of the virtual environment is shown in Table 10.12, and the corresponding relationship between the control task of the environment and the virtual control variable of the environment (EnvAgent.cs) is shown in Table 10.13.

To visually observe the entire simulation state, Unity3D must be integrated with a collaborative planning model based on MAS theory. Several common issues must be addressed, as discussed below.

Table 10.13 The corresponding table of virtual underground environment control task and virtual underground environment control task

Serial number	Virtual underground environment control task	Virtual underground environment control task
1	$M_E(i)1$: mine pressure	*YiJiaFangShi*
2	$M_E(i)2$: broken	$P_{zj}(i)$, n_{press}
3	$M_E(i)3$: accidental caving	m_{sudd}
4	$M_E(i)4$: based on user input	$H_u(m)$

10.5 VR Planning Method (FMUnitySim)

10.5.1 3D Model of the Three Machines

According to the scene mapping and a full set of drawings, the 3D model of the three machines is obtained and repaired in the UG (Unigraphics NX). Considering the capacity and pressure of the software and hardware, the external dimensions should be precisely modeled, and the internal transmission structures are ignored.

The 3D model is imported into 3DMAX in STL format and converted into FBX format, which can access Unity3D. In this manner, virtual models that are consistent with the physical model are established in Unity3D.

10.5.2 Model of the Underground Environment

The underground environment includes the intellectual input of various types of variables, such as the roof height, mine pressure and broken degree.

In this paper, to plan in a convenient manner, the floor is ideally flat, and the roof height is described with 100 points of the cutting height of the front drum, which are collected using the actual underground coal wall.

The mine pressure and broken conditions are generated in the virtual scene according to the setting result that the users input into the GUI module.

10.5.3 GUI Interface

The GUT interface can set up different initial conditions by entering different planning parameters. It is mainly divided into six modules, as shown in Fig. 10.9.

- The module of geological terrain parameters is a general overview of the underground geological environment, which includes the dip angle of the coal seam, degree of broken roof, and mine pressure regulation.
- The module of the roof and floor parameters is responsible for the roof and floor parameters, which determine the generation of the virtual roof and floor in FMUnitySim.
- The module of the coal cutting method and process parameters includes the selection of coal cutting methods and three rules between the shearer and hydraulic supports.
- The module of the shearer parameters contains the movement and performance parameters, such as the scope and acceleration of the shearer haulage speed, scope of the permitted follower distance, and virtual current of the motor.

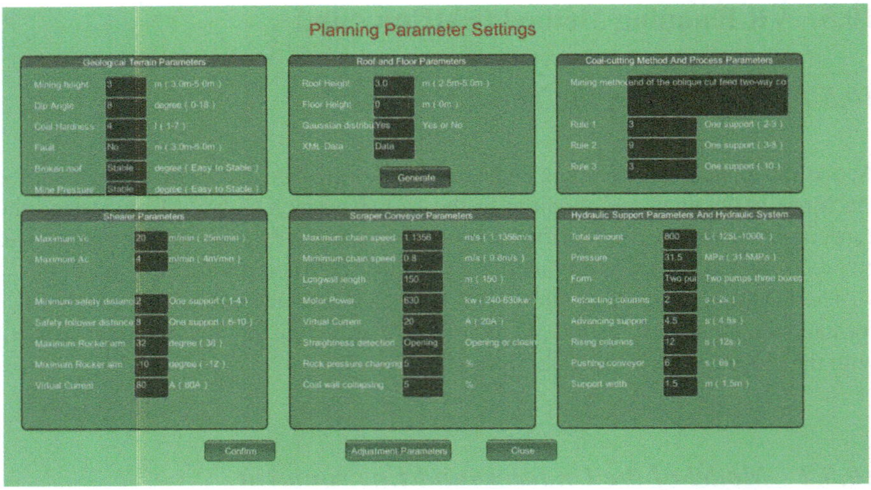

Fig. 10.9 Parameter setting interface

- The module of the scraper conveyor parameters includes the movement and performance parameters, such as the scope of the chain speed, power of the motor, detection of straightness, and collapse probability of the coal wall.
- The module of the hydraulic support parameters and hydraulic system includes the movement and performance parameters, such as the total amount, pressure and form of the hydraulic system.

The experimental planning conditions in this paper are set as follows:

- The roof is assumed to be undulating, and the floor is assumed to be flat.
- The planning process belongs to the coal cutting stage of the shearer from the scraper conveyor head to the scraper conveyor tail, and this stage is part of the end beveling feeding bidirectional coal cutting method.
- The uploading point of the scraper conveyor is assumed to be in its head section, and the direction of coal transport is assumed to be from high to low.

10.6 Summary

In this section, from the perspective of artificial intelligence and VR simulation, based on the mathematical model of 'three machines' coordination and MAS theory, a VR collaborative planning method based on MAS is proposed and a prototype system (FMUnitySim) is established under Unity3D simulation engine.According to the dynamic characteristics of the changeable environment, equipment and technology of the working face, the 'three machines' collaborative operation VR system is designed from multi-dimension by using multiple factors, and the factors affecting the coordinated operation of the 'three machines' are analyzed; Through mathematical modeling, the communication mode, coordination, conflict resolution, redundant processing perception and other issues between multiple Agents such as shearer, scraper conveyor, group hydraulic support, fluid supply system and dynamic underground environment are studied. Based on this, the "three machines" collaborative planning system is established under Unity3D, which can plan and control the key parameters of the "three machines" online. The research in this chapter provides a theoretical basis for the rapid planning and safe production of comprehensive working face.

Reference

1. Wand, Y.: Design Manual for Continuous Conveying Machinery. China Railway Publishing House, Beijing (2001)

Chapter 11
Sensor Information Architecture and Virtual Reality Interaction Technology

11.1 Introduction

In order to improve the authenticity of the constructed virtual working face, a considerable amount of sensing information is needed to drive the synchronous operation of the virtual equipment in real time. At the same time, the installation and configuration of sensors has become a major obstacle to the industrial application of virtual monitoring in fully mechanized mining.

If additional for virtual monitoring in the running of the equipment layout more sensors, is bound to make the whole production system becomes more complex, not only the rapid increasing of cost, more important is the increasing of the monitoring result in working face operation and maintenance cost is too high. The reliability of the normal operation of the system is also reduced. If the fault of the sensor causes the whole working face to shut down for maintenance, the gain outweighs the loss. At the same time, it also requires a higher professional quality of front-line operators and inspection personnel.

11.2 Real-Time and Reliable Information Acquisition and "Virtual-Real Fusion" Channel Technology

11.2.1 Sensor Placement and Perceptual Information Acquisition

Virtual monitoring system must add a minimum number of sensors to the existing electro-hydraulic control system and relatively mature configuration scheme, so as to achieve the purpose of VR monitoring. Specific sensor composition, information transmission and integration modes of each device are shown in Fig. 11.1.

X. Wang et al., *Virtual Reality Technology in Mining Machinery*,
https://doi.org/10.1007/978-981-16-4408-5_11

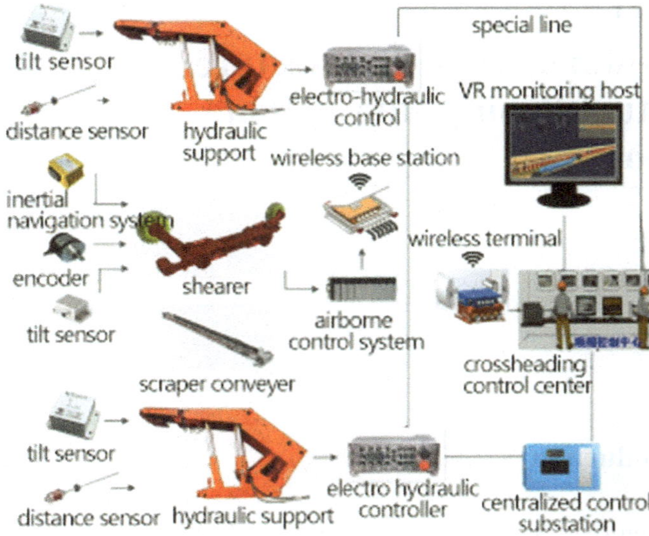

Fig. 11.1 Sensor composition, information transmission and integration of mining equipment

The position and attitude of shearer is the key monitoring point, which needs multi-source information fusion. It is necessary to carry out compound positioning of the inertial navigation system and the travel unit encoder [1] to achieve accurate positioning under the conditions of undulation and so on. Considering the vibration characteristics of shearer in the working process, the inclination sensor with good vibration resistance and high reliability should be selected to measure the shearer attitude. Each sensor and signal line is connected to the shearer airborne control system. Based on the wireless interface, it is transmitted back to the monitoring host of the centralized control center through the airborne wireless base station.

In addition to arranging strut pressure sensor and displacement sensor in the electro-hydraulic control system, it is also needs to monitor the real-time posture of the hydraulic support. Relevant inclination sensors should be arranged on the base, the four-bar linkage and the top beam. Since the movements of the support are slow, most of the time they are in a static state, and the number of support groups in the working face is also large, the inclination sensor with high cost performance can be selected after comprehensive consideration. After the sensor information of each support is integrated, it is integrated into the electro-hydraulic control system of the support, which can realize the interconnection between the supports and transmit the information back to the centralized control center.

To determine the three-dimensional morphology of the scraper conveyor, an inclination sensor should be installed on the middle pan of each section. However, this kind of sensor layout is more difficult, easy to be broken by falling coal, and its reliability is low. Therefore, based on the connection and operation relationship between the shearer and the scraper conveyor, the real-time three-dimensional morphology of the scraper conveyor can be obtained by inversion of the position and orientation obtained from the relevant sensing information carried by the shearer.

11.2.2 Key Technologies of Real-Time Interactive Channel Interface

On-line operation data of fully mechanized mining equipment are collected in real time and then enter virtual monitoring software through a series of channels. Among them, as a software platform, Unity3D can interact with configuration software, database and computing software for real-time information [2], and then drive virtual scenes for real-time synchronization, which can directly monitor the running state of the whole working face. The database stores data for monitoring and analysis, including efficient virtual memory cutting of shearer, memory posture of hydraulic supports [3] and state prediction of scraper conveyor [4], etc. Data and calculation results are transmitted and stored in the cloud. Take the signal access of Strapdown inertial navigation as an example for analysis, and the process is shown in Fig. 11.2.

The real-time communication between MTi-300 → MCU → Configuration King → Unity3D is used to realize the real-time uploading of roll angle, pitch angle and yaw angle of the inertial navigation to the upper computer. The software Configuration King and Unity3D experimental simulation platform in upper computer realize real-time data interaction through SQLsever database, and finally realize real-time interaction and control between real virtual devices on the Unity3D platform and physical devices.

Data transmission methods include RS-232 hardware protocol and Xsens company's Xbus software protocol. The serial port technology of single chip microcomputer is used to receive Xbus data packet information from the inertial navigation, and the required euler angle data information is extracted. Then use another serial port, to achieve real-time communication with King View of the upper computer.

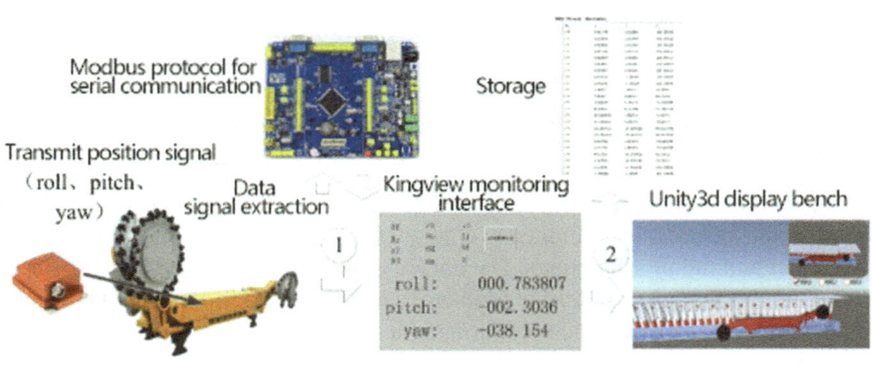

Fig. 11.2 Real time input process of sins signal

Datas are uploaded to Kingview. Kingview processes the data and pre-stores the data to the SQLsever database. Finally, using Unity3D to write C# scripts, read out the data from the pre-existing database and map it to the virtual digital model in real time.

11.2.3 Secondary Fusion and Correction of Sensing Information Data

When there is too much equipment in the working face, there will always be some problems on sensors. Therefore, in the process of overall monitoring of the working face, when data are used to directly drive the operation of virtual equipment, data jumping or wrong data will lead to sudden changes in the running state of virtual equipment, which will reduce the reliability of VR monitoring. Therefore, the virtual monitoring system must have the ability of real-time judgment, and only the data of deterministic judgment can be displayed on the virtual monitoring screen.

In order to solve the problem of data instability, the secondary information must be processed and fused. Measures are as follows:

- When selecting the sensor, in the case of the cost allows, the use of good internal anti-vibration performance sensor can filter the noise, in order to reduce the fluctuation of data signal;
- The edge calculation device is installed to carry out multi-source information fusion and transfer the secondary fusion datas, which not only reduces the pressure on the central computer, but also ensures the monitoring in the case of datas fluctuation. Because the central processing pressure is too large, it must be carried out through edge calculation. Zigbee wireless transmission method is adopted to arrange a large number of edge calculation nodes to fuse the collected data.
- The values of the previous several moments stored in the database are fused to prevent jumping;
- In the bottom layer of virtual reality software, related programs and codes are written to avoid the fluctuation of sensors.

11.2.4 Distributed Collaborative Driving Patterns

At present, the advanced explosion-proof computer equipped in the underground centralized control center is limited in configuration, and there are many monitoring equipment and monitoring points. Data processing of all equipment using a single virtual monitoring host will lead to problems such as stalling and frame rate decline, which will seriously affect the monitoring efficiency.

In order to relieve the pressure of a single host, it is necessary to carry out distributed network cooperative monitoring of multiple monitoring hosts: distribute

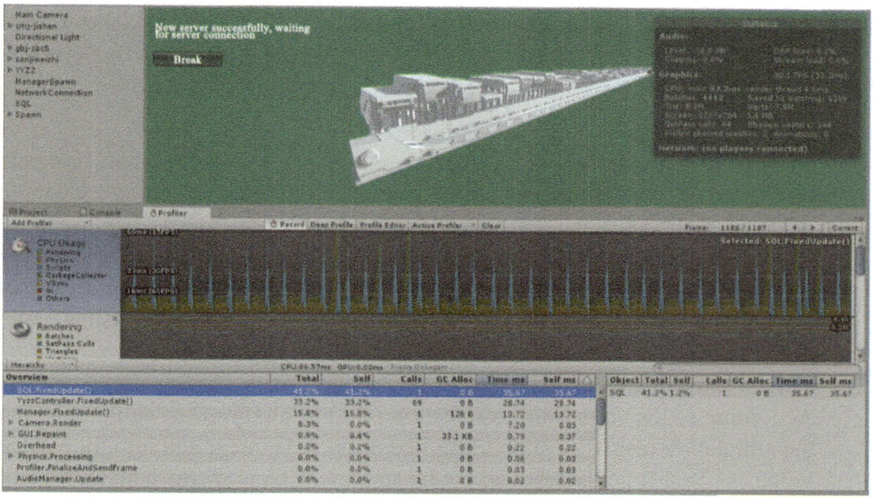

Fig. 11.3 Frame rate and memory consumption of distributed and centralized collaborative rendering

the status data of real-time operation of each equipment, so that each monitoring host can share and run synchronously in real time, reduce network pressure and accelerate data processing speed.

A distributed virtual reality monitoring system was established in Unity3D, and multiple subsystems were established. The synchronous communication program between the subsystems was written and the synchronization mode was studied. Unity3D reads the data in the database in real time and uses the data in the database to drive the model in the virtual monitor screen to take action. Multiple monitoring hosts are used to build a distributed virtual reality monitoring platform. Through the test, the collaborative comparison results between a single host and multiple hosts are shown in Fig. 11.3. The frame rate of multiple hosts is significantly reduced, information transmission is more stable, reliable and smooth, and the computer occupies less memory.

11.3 Presentation of Fusion of Virtual-Reality and Perceptual Consistency

11.3.1 Real-Time Driving Framework for Complex Fully-Mechanized Mining Virtual Scene

A virtual monitoring method driven by the fusion of sensing information and simulation information is adopted. According to the fusion results, the running state of equipment is predicted synchronously, and real-time virtual presentation is performed to improve the reliability and accuracy of monitoring, and it has the self-correction function. Different levels of driving method research are required:

(1) Single factor drive
Shearer, scraper conveyor, hydraulic support single machine and dynamic coal seam are the foundation, and real-time information driven virtual single machine method must be completed firstly. It is necessary to pay attention to the relationship between the amount of actual sensing information and the degree of freedom of virtual equipment. At this level, the secondary information that can be obtained through Sect. 11.2.3 is directly driven after processing again, as detailed in Sect. 11.3.2.

(2) Collaborative matching drive
According to the supporting connection relationship between equipment and the simulation results of supporting operation, it is integrated into the virtual scene as a real-time data-driven guidance project. Coal seam changes dynamically with the change of equipment cutting running state, and the change of coal seam gradually affects the running attitude of equipment.

(3) Virtual data interface selection
Because some sensors are difficult to arrange and cannot be installed too much on the equipment, the sensing information data is reduced compared to the degree of freedom of equipment in the virtual scene. Therefore, the virtual operation data of equipment can be determined on a large scale according to the virtual contact model. However, the operating state of equipment determined by virtual simulation is sometimes contradictory with the real-time sensing information. How to solve this contradiction by composite fusion driving is also a key problem.

11.3.2 Key Technologies of Driving Virtual Stand-Alone Support Based on Underlying Model

The sensor information data cannot directly drive the virtual model to reflect its corresponding state in real time. The reason is that there is a parent–child relationship between equipment in the virtual reality scene. The movements of the parent object affects the movements of the child object, and the child object follows the movements of the parent object. Take the equipment inclination angle as an example to illustrate: the angle information driven by each component should be the value of the relative change between the parent object of the upper stage. The value of the sensor is the real value relative to the time coordinate system, which must be converted from angle to drive the operation of the virtual object.

In order to monitor the height of shearer drum, it is necessary to calculate the value of the inclination sensor on the rocker arm with the three-dimensional attitude angle of the body, so as to find out the true angle of the rocker arm. The hydraulic support must also be based on the angle of the base, the four-bar linkage, shield support and

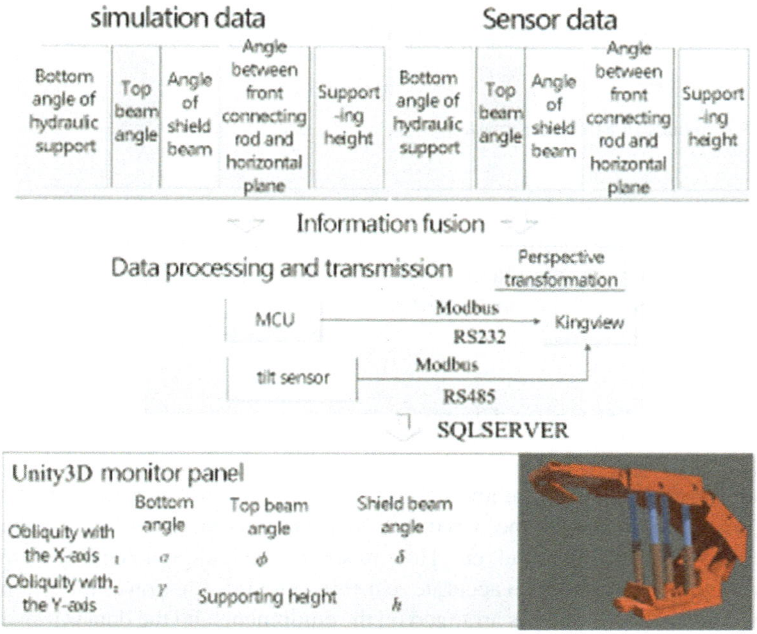

Fig. 11.4 Key technologies of driving virtual hydraulic support

other real-time conversion, in order to obtain accurate posture. And the equipment virtual simulation data is fused to improve the accuracy of virtual rendering, as shown in Fig. 11.4.

11.4 Transparent Fully Mechanized Mining Face Industrial Internet Infrastructure Construction

11.4.1 Wired/Wireless Full Coverage Network Communication Platform

The communication network platform combining wired communication and wireless coverage was established, and the gigabit industrial ring network switch and wireless WIFI network coverage were adopted. Two mine flameproof network switches are installed in the centralized control center as the data core switches, which are used to transmit all the data of the centralized control system of the working face and shoot the data back to the monitoring host in the remote dispatching center. A mine intrinsical-safety base station KT113-F is installed on the head and tail of the scraper conveyor respectively, which is used to transmit the field equipment data

Fig. 11.5 Principle of SINS transmission

to the channel centralized control center. To realize the reliable communication of various mobile devices in the underground working face environment, the video, voice and data networks are integrated into one, and a high-speed industrial ethernet communication network is constructed.

11.4.2 Digital Sensing Element

Because fully mechanized mining equipment is always in the process of dynamic migration and movements, the wired sensor installed is prone to line winding and being damaged by falling coal, etc. How to set the basic digital sensing element is very critical. In order to obtain accurate real-time pose information of the equipment, the position and pose sensor is arranged on the equipment, and the data is transmitted through the shearer airborne base station and the hydraulic support electro-hydraulic control system. The most critical technical problems include two parts:

- The design of stable and reliable wireless sensor to carry out the morphology detection of scraper conveyor;
- There is a lot of information about Strapdown Inertial Navigation System (SINS) arranged on the shearer body, so how to transmit it accurately and reliably through wireless mode.

The principle of SINS transmission is shown in Fig. 11.5.

11.4.3 Remote Monitoring System

This part includes the simulated cachet centralized control center and the simulated remote dispatching center located in the laboratory building. Among them: a core switch, six main controlling computers, a set of PLC control system (complete system chain control and data exchange between the equipment), a support intrinsic safety work station, and a shearer intrinsic safety work station, 6 intrinsic safety display screens and 7 pieces of DC12V switching power supply were equipped in centralized control center, and on the platform beside 2 explosion-proof computers configuration, respectively used for shearer remote configuration control and support electro-hydraulic control system of the system parameter setting and selection function.

At the completion of shearer remote monitoring on the basis of single set control center was also able to individual machines or other subsystem signal acquisition, comprehensive coordination, according to the process, supporting real-time operating condition and emergency changes, such as real-time adjustment of compressors operating conditions, complete the automatic coal mining and collaborative transportation under different operation conditions.

11.5 Development of Application Layer of Transparent Fully Mechanized Mining Face System

11.5.1 Two-Dimensional Data Monitoring

As the intermediate means of the whole system operation, Kingview undertakes the tasks including: data monitoring, data regularization, remote monitoring and operation, and database saving connection. As a normal upper computer software to complete data monitoring.

11.5.2 Video Monitoring

As shown in Fig. 11.6, a flameproof camera is installed in the head and tail of the scraper conveyor, and a camera is installed in the area where the equipment is running, so as to realize the monitoring of the entire working area of the working face. The collected image data is connected to the nearby switch through the wire of the flameproof network, in which the nose switch is connected to the 3 flameproof cameras, and the tail is connected to the 4 flameproof cameras. Every five sets of hydraulic support in working face to install a video monitoring system, through the integrated access device access communication network, according to the position and posture of the shearer, video can switch to camera video screen, within the scope of recently completed visual management of the whole mining equipment, mainly based on the ASC II communication of remote control switch function of the camera.

11.5.3 VR Integrated Attitude Monitoring and Early Warning System

To realize the operation state monitoring of equipment, establish the virtual image of the real fully-mechanized mining working face, and drive the virtual scene to synchronize, so as to directly monitor the operation state of the whole working face, The following problems need to be solved:

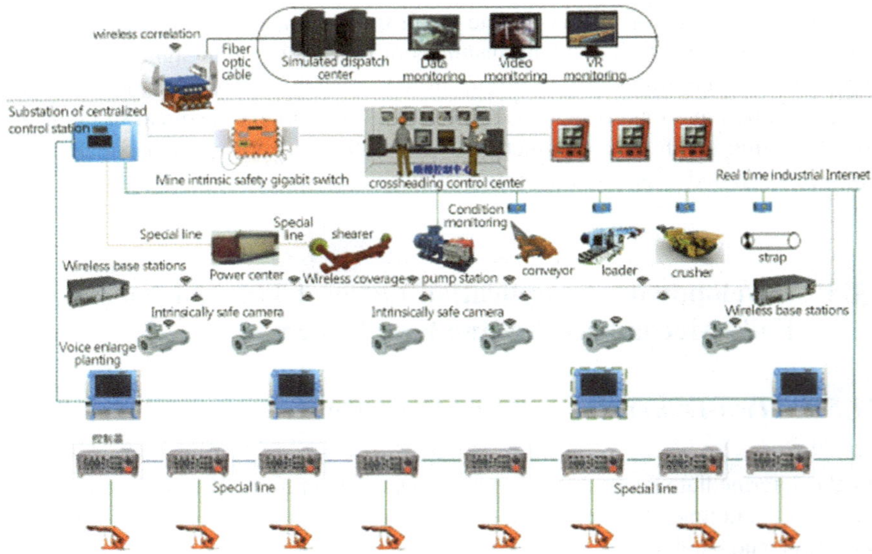

Fig. 11.6 Overall structure of transparent fully mechanized mining face

(1) Real-time construction technology of high reliability panoramic virtual reality working face scene
The Kingview monitoring host of the centralized control center saves the online data of sensor components in the SQLSever database in real time. Through the interface between the database written and C#, Unity3D access and read the online information of all corresponding sensors in real time [5]. Then, the latest data read in real time is assigned to the relevant variable interface of the corresponding virtual equipment to realize the real driver of each virtual equipment. Finally, the online real driver of the virtual scene of the fully-mechanized mining face is realized, as shown in Fig. 11.7.

(2) Fast positioning technology of mechanized mining virtual scene and video monitoring linkage
Set up the part in the virtual scene and virtual camera position corresponding to the actual video monitoring, through the sensor information of decision, determine the problem and the location of a danger happening, in the upper right corner of the whole virtual monitor screen as a virtual camera display position, quickly switch parts and locking the virtual camera of the location, to help operators find out the reason and implement remote human intervention operation.

(3) Distributed 3D virtual scene remote monitoring technology
Based on the NetworkView component of Unity3D and remote data simulation technology, the LAN (Local Area Network) in VR environment is coordinated and synchronized, the data is distributed, and then a complete virtual picture is synthesized, so as to reduce the operating pressure of the system [6].

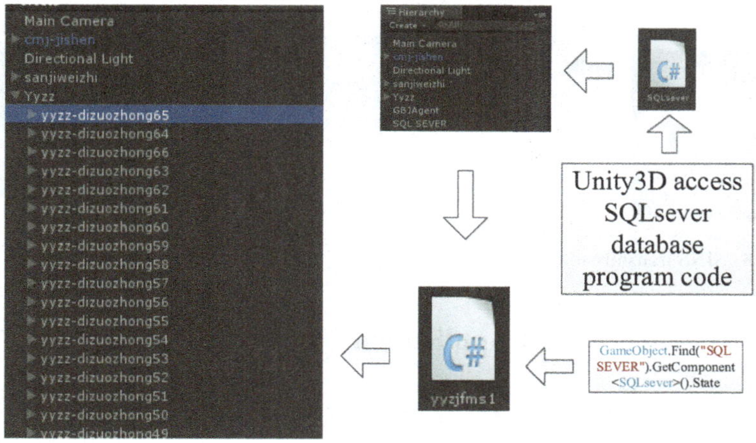

Fig. 11.7 Schematic diagram of connection between data and Unity3D software

(4) Attitude detection and early warning system
By detecting the attitude parameter data of each equipment, the attitude changes of each equipment in the operation process can be calculated according to the geometric size and matching relationship of the equipment, so as to detect the possible interference phenomenon between the equipment in advance. The virtual screen can calculate the process parameters in real time and display the equipment operation problems in stereo according to the background decision-making.

11.5.4 Three-Dimensional Transparent Monitoring System Integration

The work is carried out on the fully-mechanized mining remote control platform that integrates VR monitoring and underground video. The contents of the six display screens in the centralized control center are as follows: 1# screen: display bracket video monitoring; 2# screen: display shearer body dust removal camera; 3# screen: display video images of fixed points; 4# screen: display shearer data and support electro-hydraulic control system information; 5# screen: display VR surveillance screen 1; 6# screen: display VR surveillance screen 2.

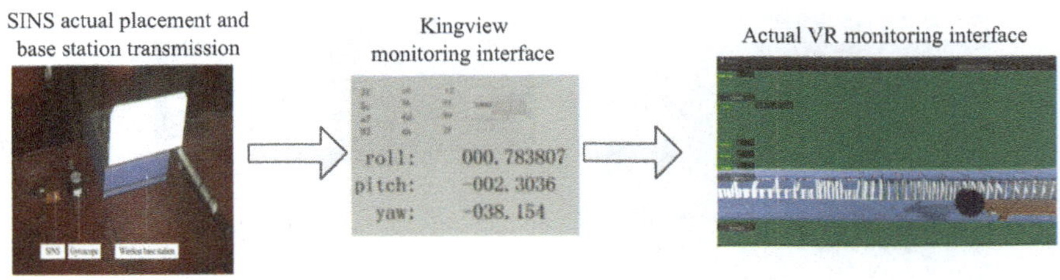

Fig. 11.8 SINS transmission test

11.6 System Test and Application

11.6.1 Digital Sensing Element Testing

According to the layout plan of digital sensing element, each element is installed on each equipment. Taking the coal shearer INS transmission as an example to carry out the test, the information transmission is normal, can be seamlessly transmitted back to the Kingview software of centralized control center, and can be smoothly connected with the virtual screen, the operation is normal, as shown in Fig. 11.8.

11.6.2 Network Performance and Functional Testing

After the debugging of each subsystem is completed, the whole system is integrated. Firstly, the performance test of the whole industrial Internet is carried out, and the virtual sensing and video monitoring test is taken as an example by installing the corresponding digital sensing components on a single hydraulic support. Tests were carried out in the simulated centralized control center and the simulated remote dispatching room respectively, and the results were shown in Fig. 11.9. The correctness of the virtual bracket motion and the delay of the motion data were counted. In the simulation centralized control center, the screen runs smoothly, and actual bracket movements can basically keep synchronization. Through the proprietary test

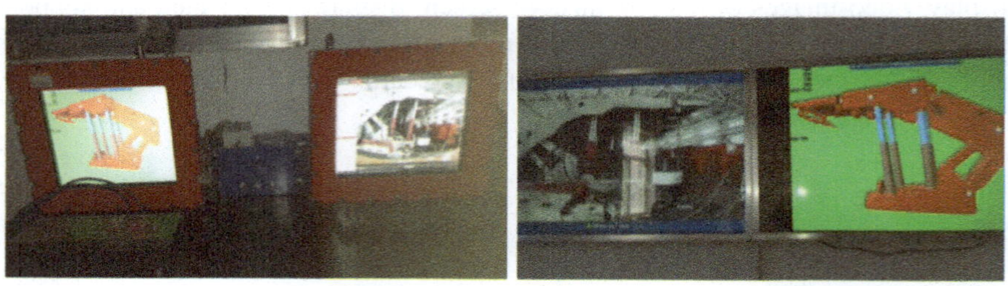

Fig. 11.9 Centralized control center test and remote dispatch center test

system, it is concluded that the action delay is about 300 ms, while the stability of video monitoring is poor, and the delay is 0.5–2 s. When the same experiment was carried out in the simulated remote dispatching center, the delay of both methods increased slightly. It shows that VR monitoring has advantages over video monitoring in picture delay and signal transmission and can be used as a good supplement.

11.6.3 Visualized Coal Mining Experiment with Remote Intervention

Based on the omnidirectional stereo monitoring mode and system, the remote visualization coal mining and VR monitoring tests were carried out in the centralized control center and remote dispatching room along the chute. The PLC control element is installed in the shearer control box to supply power to the high voltage circuit and control loop. After the communication between the upper computer and the field equipment, the upper computer can control the field equipment. Two cutting drums of the shearer up and down can be controlled by the upper machine in the left up, left down, right up and right down green arrow button, shown in Figs. 11.10 and 11.11. The test shows that the operation response time of the test remote control is less than 200 ms. Shearer alarm trigger emergency stop time: less than 100 ms, to meet the requirements of remote control precision.

Fig. 11.10 Using the physical "three machine" test, remote visualization coal mining

Fig. 11.11 Remote visual coal mining operation interface

11.6.4 Coordinated Operation Test of Three Engines

The coordinated operation experiment of three machines was carried out in the centralized control center. The signal of the shearer, hydraulic support groups, and scraper conveyor was transmitted back to the centralized control center in real time, and the automatic control program written by the centralized control system was experimented.

11.7 Conclusion

This chapter introduces the sensor information construction technology and virtual and real interaction technology. It introduces the real-time reliable information acquisition and virtual-real fusion channel technology from four aspects: sensor placement and perceptual information acquisition, key technologies of real-time interactive channel interface, the secondary fusion of sensing information data and the modified driver mode of distributed collaboration. The virtual-real fusion and perceptual consistency presentation methods were studied from two aspects: real-time driving framework of complex fully-mechanized mining virtual scene and the key technology of the underlying model driving virtual single machine. It introduces the construction of industrial Internet infrastructure of transparent fully mechanized mining face from three aspects: wired and wireless full coverage network communication platform, digital sensing original and remote monitoring system. It introduces the development process of application layer of fully mechanized mining face system from four aspects: two-dimensional data monitoring, video monitoring, VR integrated attitude monitoring and early warning system, three-dimensional transparent monitoring system integrated transparency. Finally, the system test and application are carried out.

References

1. Li, A., Hao, S., Wang, S., et al.: (2016) Experimental study on shearer positioning method based on SINS and encoder. Coal Sci. Technol. **4**, 95–100 (2016)
2. Xie, J.: Method of Monitoring and Dynamic Planning for Three Machines in a Fully Mechanized Coal-mining Face under VR Environment, p. 2018. Taiyuan University of Technology, Taiyuan (2018)
3. Xie, J., Wang, X., Yang, Z., et al.: Attitude-aware method for hydraulic support groups in a virtual reality environment. Proc. Inst. Mech. Eng. C J. Mech. Eng. Sci. **233**(14), 4805–4818 (2019)
4. Qiao, C.G.: Research on the theory and method of coordinated position monitoring for shearer and scraper conveyor. Taiyuan University of Technology, Taiyuan (2019). (in Chinese)

5. Xie, J.C.: Method of monitoring and dynamic planning for three machines in a fully mechanized coal-mining face under VR environment. Taiyuan University of Technology, Taiyuan (2018). (in Chinese)
6. Shi, H., Xie, J., Wang, X., et al.: An operation optimization method of a fully mechanized coal mining face based on semi-physical virtual simulation. Int J Coal Sci Technol. **7**, 147–163 (2020)

Chapter 12
Working Condition Monitoring and Virtual Simulation Method

12.1 Introduction

Based on the research of single machine working condition monitoring and virtual simulation method, it is necessary to study the attitude behavior of Three machines in fully mechanized mining face under the actual working condition. The virtual simulation method can simulate many situations and effectively support the "three machine" attitude monitoring method. Therefore, this chapter compiles the mathematical model of the "three machine" actual attitude behavior in VR environment, meanwhile, the compiled results can reverse verify and support the actual "three machine" condition monitoring method.

12.2 "Three Machine" Virtual Co-simulation Under the Condition of Horizontal Ideal Floor

12.2.1 Key Technologies of Shearer Virtual Walking

12.2.1.1 Change and Analysis of Shearer Walking Path

During the bending process of scraper conveyor, all pin rows and pin shafts are connected in sequence as the virtual path of shearer, as shown in Fig. 12.1.

12.2.1.2 Setting Approximate Virtual Path of Shearer

The left and right pin rows and pin shafts corresponding to each middle slot are set as the waypoint attribute. The parent waypoint and child waypoint can be set for each waypoint, corresponding to m_p and m_n respectively. Manual assignment is required,

X. Wang et al., *Virtual Reality Technology in Mining Machinery*,
https://doi.org/10.1007/978-981-16-4408-5_12

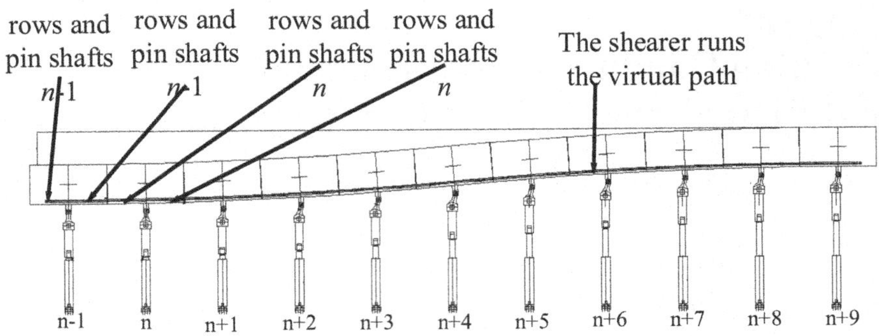

Fig. 12.1 Virtual motion trajectory of shearer motion

Table 12.1 Waypoint, parent and child waypoint settings

Waypoint	Parent waypoint (m_p)	Child waypoint (m_n)
Left pin row and pin shaft $n-1$	Right pin row and pin shaft $n-2$	Right pin row and pin shaft $n-1$
Right pin row and pin shaft $n-1$	Left pin row and pin shaft $n-1$	Left pin row and pin shaft n
Left pin row and pin shaft n	Right pin row and pin shaft $n-1$	Right pin row and pin shaft n
Right pin row and pin shaft n	Left pin row and pin shaft n	Left pin row and pin shaft $n+1$
Left pin row and pin shaft $n+1$	Right pin row and pin shaft n	Right pin row and pin shaft $n+1$
Right pin row and pin shaft $n+1$	Left pin row and pin shaft $n+1$	Left pin row and pin shaft $n+2$

such as the waypoints of middle slot *n-1*, middle slot *n* and middle slot$n+1$, as shown in Table 12.1.

Assign waypoints one by one, and use the command SetNext (PathNode node) to set the parent and child waypoints of each waypoint. When the running direction of the shearer is overturned, the reverse setting of each road point is carried out again.

12.2.1.3 Key Technology of Shearer Walking Along Virtual Path

Add to Shearer lujing.cs Script, and set the current waypoint *nc* as the first waypoint (scraper conveyor head waypoint), the shearer automatically searches the virtual path of shearer to move.

Initial alignment: In order to drive the Shearer to move along the virtual path, a virtual parent object is set up on the top level of the highest level of the shearer model, which is consistent with the size of the waypoint object and in a straight line with the coordinates of the initial waypoint. The virtual object mainly controls the movement of the shearer through RotateTo() and MoveTo().

RotateTo (): It means that the angle between the current position and the first waypoint picked up by the shearer and corrected; when the process section is converted, the object marked by the shearer position needs to be turned 180 degrees;

MoveTo (): It means that the shearer is pulling along the direction of turning.

12.2.2 Mutual Perception Technology Between Shearer and Hydraulic Support

12.2.2.1 Coupling Strategy of Shearer Speed and Hydraulic Support Groups Action

Because the pushing and sliding speed of the hydraulic support is faster than the hauling speed of the shearer and the moving speed of the hydraulic support, only the coordination of the hauling speed V_c of the shearer and the moving speed V_y of the hydraulic support is needed to better grasp the automatic operation relationship of the Three machines in fully mechanized mining face.

Among them:

$$V_y = \frac{D_{zj}}{t_1 + t_2 + t_3} \tag{12.1}$$

In the formula: t_1: Column falling time; t_2: Moving time; t_3: Pushing and advancing time. This book takes the hydraulic support 200 mm away from the roof as an example to calculate.

When $V_c < V_y$, the support can be moved according to the order of follow-up, and the effect of follow-up is good, and there will be no problems such as losing the rack or not moving the support in place. Before the shearer position triggers the next group of supports to move with the machine, the previous group of supports has completed the automatic moving with the shearer, which can meet the requirements of the orderly moving with the shearer in the fully mechanized working face.

When $V_y < V_c < 2V_y$, the way that the support moves with the shearer in sequence cannot meet the requirements of the support tracking shearer. It can follow the shearer to move the support in sections or insert the support into multiple supports to move the support, and the goal can be achieved by moving multiple supports at the same time. The first, third and fifth supports can be moved at the same time, and then the second, fourth and sixth supports can be triggered to move at the same time, which greatly improves the moving speed.

12.2.2.2 Mutual Perception Between Shearer and Hydraulic Support

Every hydraulic support is controlled by the YyzzControl.cs Script, every shearer is controlled by script CmjControl.cs Script. The perception of shearer and hydraulic support is mainly carried out through the following three rules:

- Rule 1: When the support falls behind the back drum of the shearer by two, it begins to perform the action of lowering-moving-raising;
- Rule 2: When the support is 10–15 m behind the shearer, it begins to push and slide;
- Rule 3: When the support is ahead of the two supports of the front drum of the shearer, it begins to retract the guard board;

The above describes the way to move the support in order.

12.2.3 Mutual Perception Technology Between Hydraulic Supports

The hydraulic support needs to have the ability to sense the movement of the support in a certain range around. When moving the support in sequence, and the hydraulic support has been activated at the position of the back drum of the shearer, the hydraulic support also needs to sense whether the previous support has been moved or not. If the previous support is still moving, it needs to wait for the previous support to complete the movement before starting the movement. The code is as follows:

```
If ((HouGunTongWeiZhi-cmj.transform.position > 2*Dzj)&&(GameObject.
Find (NextID(YyzzID)).GetComponent < YyzzFMS > ().State = = 5)&&(YiJia
= = false)).
{State = 2; YiJia = true;}
```

When multiple supports move at the same time, the sensing range of support needs to be expanded. For example, when two supports move at the same time, the sensing range of supports needs to be set to 3 to sense the supports with a long distance.

12.2.4 Mutual Perception Technology of Hydraulic Support and Scraper Conveyor

The mutual perception between the hydraulic support and the scraper conveyor is mainly the virtual simulation of the pushing process between the hydraulic support pushing cylinder and the middle trough.

Because the pushing mechanism is a floating mechanism, it has a certain degree of freedom. The push rod can be bent adaptively; the pin axis coordinates of the push cylinder and the initial push–pull hole of the bracket can be calculated respectively, and the coordinates between the two can be obtained in real time and calculated, so as to calculate the extension length of the push cylinder and other parameters in real time.

12.2.5 *Mutual Perception Technology of Virtual Three Machines in Fully Mechanized Mining Face and Coal Mining Technology*

Different coal mining methods have different effects on the simulation of fully mechanized mining technology. Take the end oblique cutting and feeding, for example, it can be divided into three sections and six stages.

Three sections: head section, middle section and end section;

Six stages: Head oblique cutting and feeding, head cutting the triangle coal, normal cutting from the head to the end, end oblique cutting and feeding, end cutting the triangle coal and normal cutting from the end to the head;

Eight parameters:

C1: The distance that the shearer pushes the scraper conveyor forward; C2: The distance that the shearer pushes the scraper conveyor back; P1: The distance between front roller and center; P2: The distance between rear roller and center; Q: Safe distance; W: Length of bending section; M: Working face length; A: Length of the head hydraulic support.

Head part: Ensure that the head support to the last support in the bending section are moved and pushed while the shearer machine is oblique cutting and feeding.

Parameter solution: According to the set parameters, determine the range of the three section respectively.

$M = 100$; $A = 3$; $P_1 = 3$; $P_2 = 5$; $C_1 = (6–10)$; $C_2 = (2–3)$; $Q = 4$; $W = 9$; The length of the head section and end section can be calculated as follows: $(2–3) + 8 + 4 + 9 + (6–10) + 8 = (38–42)$.

12.2.6 *Principle of Consistency in Time and Unit*

Functions such as Start() and Update() are used to write programs in Unity3D. Updata() is called every time a new frame is rendered. It runs at different speeds under different computer configurations. FixedUpdate() is executed at fixed intervals, independent of frame rate. So use FixedUpdate() for event updates.

Where the FixedUpdate() interval can be changed in the project settings by successively manipulating Edit → Project Setting → time, and then finding the Fixed Timestep option to set the value. This section is set to 0.2 s.

Table 12.2 Calculation of frame moving time of ZZ4000/18/38

Column dropping (mm)	Column dropping (s)	Frame moving (s)	Column rising (s)	Total time (s)	The corresponding V_c (m/min)
200	1.74	4.36	12.06	18.16	4.96
100	0.87	4.36	6.03	11.26	7.99

Units in Unity3D: actually 1 m = 10 units in Unity3D (so equivalent to dm). Take the speed of support movement as an example to illustrate the principle of consistency in time unit, as shown in Table 12.2.

12.3 Coupling Method of Cutter Attitude Between Shearer and Scraper Conveyor

Aimed at the important problem that the scraper conveyor should have a good posture of "S shape" in the bending process, a method of calculating the posture of the bending section chute by analyzing the key dimension coordinates is studied. This method modifies and improves the existing solution method of bending section, and can precisely solve the expansion length of the bending section chute, the pin row and the push cylinder corresponding to each section of the chute. Then the relationship between the change of the yaw angle and the path of the shearer in the process of passing through the S-shaped bending section can be studied.

12.3.1 Solving Calculation Process Model of Bending Section

12.3.1.1 Solve the Modified Calculation Model for Bending Section

Let θ_n be the angle between the n_{th} curved section chute and the coal wall line in one side. It can be concluded that $\theta_1 = \alpha$, $\theta_2 = 2\alpha \ldots \theta_n = n\alpha$. According to the symmetry and force analysis, when the total number of chute in bending section is odd, the number of chute sections in bending section is the $2n - 1$th chute, and the symmetry center is the center of plate in the Nth chute (Fig. 12.2).

After finishing, the number of chute segments N in the unilateral bending section can be obtained:

$$N = \frac{1}{\alpha}\cos^{-1}\left[\cos\frac{1}{2}\alpha - \frac{(B + a - a\cos N\alpha + L\sin N\alpha)\sin\frac{1}{2}\alpha}{L + b}\right] - \frac{1}{2} \quad (12.2)$$

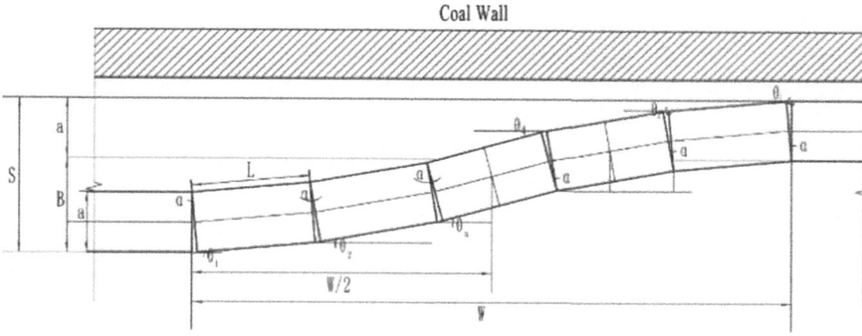

Fig. 12.2 Schematic diagram of odd segment calculation

12.3.1.2 Determination of Na Value

N_a is the number of one-sided middle slot segments in the precise bending symmetry interval, which is determined according to the following principles:

- If the degree of α is small, Na can choose the largest positive integer less than N;
- If the degree of α is large, Na can choose the smallest positive integer greater than N.

However, when calculating the central coordinates of the bending section, the Y coordinate of the central point will exceed the Y coordinate of the central chute of the NA section under some circumstances, which needs to be further corrected.

12.3.1.3 Accurate Determination of the Exact Bending Angle

By substituting the N_a value into the following equation, the exact bending Angle α_a can be obtained.

$$S = 2(L + b)\sum \sin N_a\alpha_a + a\cos N_a\alpha_a - L\sin N_a\alpha_a \tag{12.3}$$

12.3.2 Solution of Bending Section Chute Attitude

Assuming that the bottom plate is flat and there is no longitudinal bending, at this time, only the central coordinates of the plate in the chute and the deflection angle of the chute are needed to determine the attitude of the chute.

12.3.2.1 Chute Attitude Solution

Taking the shearer cutting from right to left, for example, the middle plate center of each bending section of the chute is selected as the center of each chute, and the lower right vertices of the chute (the first chute on the left in Fig. 12.2) which is about to enter but has not entered the state of pushing and sliding are set as the origin of coordinates. Then the central coordinates of the i_{th} chute of the bending section (the first chute on the left of the bending section is numbered 1) can be expressed as the following form: (X_i, Y_i, θ_i). The three variables can be expressed as the following piecewise function:

$$X_i = \begin{cases} \left(b+\frac{L}{2}\right)\cos\theta_i - \frac{a\sin\theta_i}{2} & i=1 \\ \sum\limits_{j=1}^{i-1}(L+b)\cos\theta_j + \left(b+\frac{L}{2}\right)\cos\theta_i - \frac{a\sin\theta_i}{2} & 2\le i\le N \\ \sum\limits_{j=1}^{N}(L+b)\cos\theta_j + \frac{L}{2}\cos\theta_i - \frac{a\sin\theta_i}{2} & i=N+1 \\ \sum\limits_{j=1}^{N}(L+b)\cos\theta_j + \sum\limits_{j=N+1}^{i-1}L\cos\theta_j + \frac{L}{2}\cos\theta_i - \frac{a\sin\theta_i}{2} & N+1<i\le 2N-1 \end{cases} \tag{12.4}$$

$$X_i = \begin{cases} \left(b+\frac{L}{2}\right)\cos\theta_i - \frac{a\sin\theta_i}{2} & i=1 \\ \sum\limits_{j=1}^{i-1}(L+b)\cos\theta_j + \left(b+\frac{L}{2}\right)\cos\theta_i - \frac{a\sin\theta_i}{2} & 2\le i\le N \\ \sum\limits_{j=1}^{N}(L+b)\cos\theta_j + \frac{L}{2}\cos\theta_i - \frac{a\sin\theta_i}{2} & i=N+1 \\ \sum\limits_{j=1}^{N}(L+b)\cos\theta_j + \sum\limits_{j=N+1}^{i-1}L\cos\theta_j + \frac{L}{2}\cos\theta_i - \frac{a\sin\theta_i}{2} & N+1<i\le 2N-1 \end{cases} \tag{12.5}$$

$$\theta_i = \begin{cases} 0 & i\le 0 \\ i\alpha & 0<i\le N_a \\ (2N_a-n)\alpha & N_a<i\le 2N_a-1 \\ 0 & 2N_a-1<i \end{cases} \tag{12.6}$$

12.3.2.2 Calculation of Middle Trough Structure

Figure 12.3 is the calculation diagram of the chute coordinate. The center of the chute plate is selected as the central point of the chute, where: G_1: Horizontal distance between chute centerline and pin-row centerline; G_2: Horizontal distance between the chute centerline and the chute push–pull hole centerline; Lg_1: the distance between the center line of the chute and the center line of the axle hole of the middle pin row

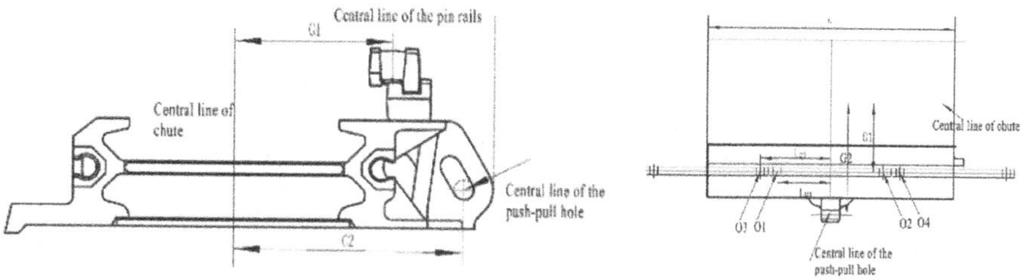

Fig. 12.3 Calculation diagram of chute coordinates

of the chute; Lg_2: the distance between the center line of the chute and the center line of the shaft hole of the chute connecting pin row;

The analysis shows that: $Lg_2 = L/2 - Lg_1$.

12.3.3 Update and Calculation of Shearer's Walking Path

12.3.3.1 Walking Model of Shearer

Shearer walking depends on the traction part of the walking wheel and connected in the chute on the pin row meshing to walk. The meshing between the walking wheel and the pin row of the hauling part of the shearer is similar to the principle of pinion and rack, in which there is a guide sleeve between the two to guide, in which the guide sleeve guide width > Exclusive width > Width of pin row teeth > the width of the walking wheel, which enables the whole front and rear walking wheel to adapt to the small bending of the pin row in the walking track to move forward safely and smoothly.

12.3.3.2 Coordinate Analysis of Chute Pin Row

The pin row is divided into the chute middle pin row and the chute connecting pin row (Fig. 12.3), which are connected with the chute pin row base through the left and right pin shafts. The middle pin row moves along with the chute as a whole, and its relative position with the center of each chute remains unchanged, while the connecting pin row changes correspondingly with the bending degree of two adjacent chutes.

After calculation, there are $4Na - 1$ pin rows, including $2Na - 1$ intermediate pin rows and $2Na$ connecting pin rows.

The coordinate of each axle hole can be obtained respectively, and then the pin arrangement curve can be obtained by fitting.

Wherein, the axis whole coordinates on the right side of the chute middle pin row are:

$$\left(X_i + \sqrt{(L_{g1})^2 + (G_1)^2}\sin\left(\arctan\frac{L_{g1}}{G_1} + \theta_i\right),\right.$$
$$\left.Y_i - \sqrt{(L_{g1})^2 + (G_1)^2}\cos\left(\arctan\frac{L_{g1}}{G_1} + \theta_i\right)\right) \tag{12.7}$$

The coordinates of the left axle hole of the chute middle pin row, the coordinates of the left axle hole of the chute connecting the middle pin row and the coordinates of the right axle hole of the chute connecting the pin row can all be expressed in the similar form as above, which will not be repeated here.

12.3.3.3 Analysis of Chute Pin Row Equation

Bending Angle of the middle pin row:

$$\theta_{\mathrm{Mi}} = \theta_i \tag{12.8}$$

Bending Angle of connecting pin row:

$$\theta_{\mathrm{Ci}} = (\theta_i + \theta_{i-1})/2 \tag{12.9}$$

Combined with the coordinates of each shaft hole, the curve equation of pin arrangement can be expressed.

12.3.3.4 Shearer Feed Drift Angle Calculation Method

After getting the equation of pin row curve, it is necessary to study the drift angle of shearer body in the process of cutting. The shearer feeding process is mainly affected by the coupling effect of two travelling wheels and guide slipper and pin row track, which leads to the change of the shearer fuselage yaw Angle, as shown in Fig. 12.4.

Left wheel position as feature point position of shearer, and then according to the fuselage length in pin on a path to find the right position, walking round about

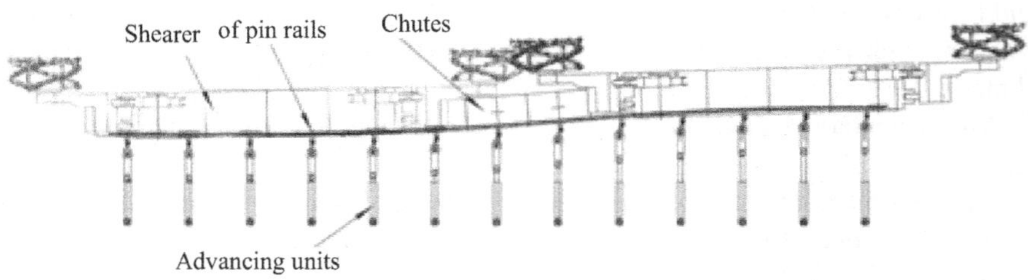

Fig. 12.4 Variation of feed yaw angle of shearer

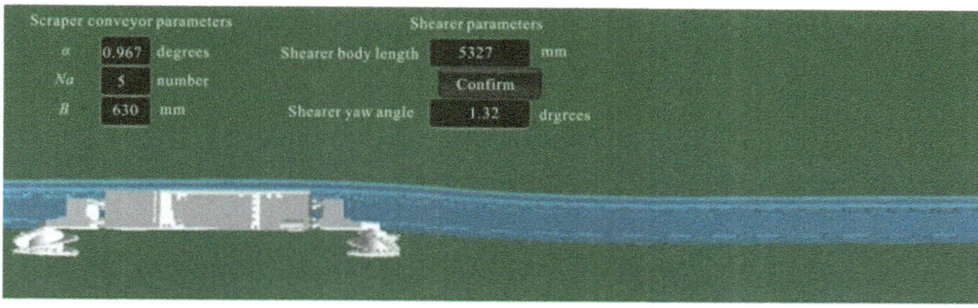

Fig. 12.5 Unity3D software

walking wheel position is connected with the angle of the lateral line drift angle of shearer Ψ_{js}, yaw angle and the corresponding to the left and right sides of shearer walking round the angle pin row of the poor is respectively coal winning machine walking wheel and pin angle difference, type can be concluded as follows:

$$\psi_{js} = \tan(Y_{A2} - Y_{A1})/(X_{A2} - X_{A1}) \tag{12.10}$$

Among them, point A_1 and point A_2 are the key points of the left and right walking wheels respectively.

According to the body length of a series of shearer matched with the scraper conveyor, the theoretical research is carried out respectively. The shearer's body length is 4500, 4900, 5327, 5800 and 6300 mm in turn.

In Unity3D, the algorithm in this section is programmed into the background program to generate a VR planning software for coupling shearer yaw Angle and scraper conveyor shape. After compiling and publishing, the length of the fuselage is changed respectively for virtual simulation, and the process data is output to XML file in real time for data analysis. The interface of virtual VR planning software is shown in Fig. 12.5, and the simulation results are shown in Fig. 12.6.

The simulation results show that:

(1) With the increase of the length of shearer body, the value of the maximum yaw angle of shearer increases with the same "S-shaped" shape of scraper conveyor;
(2) with the shearer left walking wheel as the shearer position location point, in the first half cycle, in the same scraper conveyor "S shape" shape conditions, whenever the shearer in the same position of the scraper conveyor, the longer the shearer body, the greater the corresponding deviation angle. The second half of the cycle is the opposite;
(3) To judge the bad force condition by judging the relative angle between the left and right walking wheels and the track of the pin row. If the relative Angle is larger, the force condition is worse. The smaller the relative Angle, the more normal the force. The results are shown in Fig. 12.7. It can be seen that the trajectory of the left walking wheel coincides with the shearer's deviation angle before the coordinate is 0 mm, while that of the right walking wheel coincides with the shearer's deviation angle after 8500 mm. The reason is that these

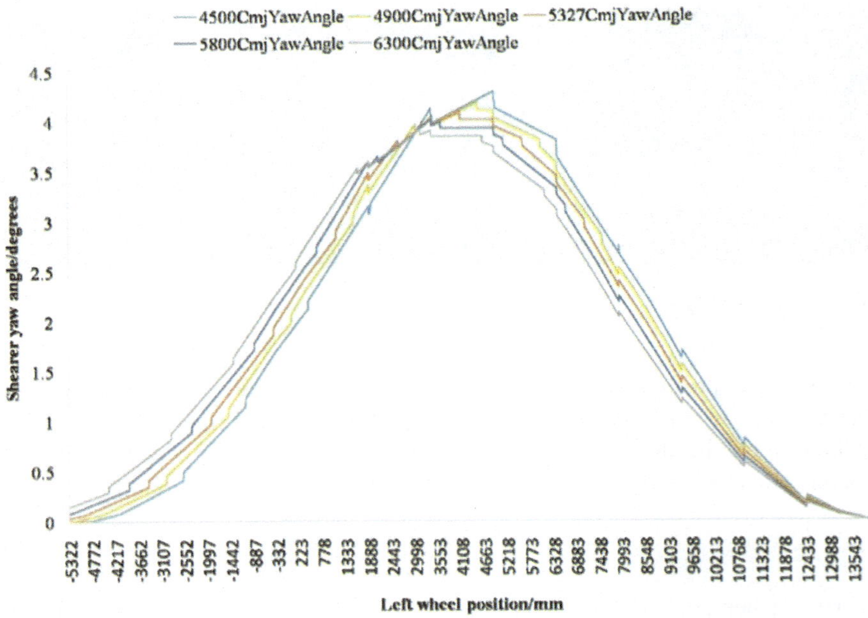

Fig. 12.6 Yaw angle of shearer inclined cutting tool under different lengths

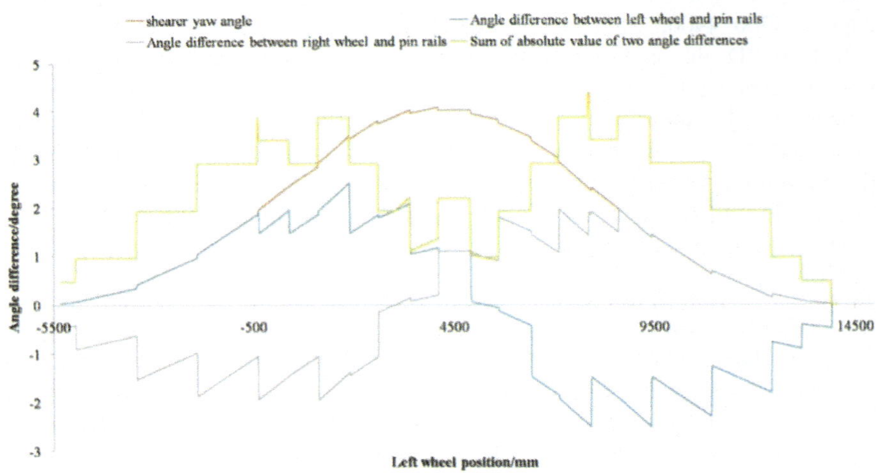

Fig. 12.7 Relative angle change trend between left and right walking wheel and pin row track

two stages are the stage where the left travelling wheel has not entered the "S-shaped" bending section and the stage where the right travelling wheel has exited the "S-shaped" bending section. In these two stages, the bending Angle of the pin row at the position of the left and right walking wheels is 0;

(4) According to the comprehensive effect of the two walking wheels, the force trend is gradually increasing in the first half, then decreasing slightly in the middle part, then rising again in the second half, and finally decreasing gradually with the operation of the shearer.

12.3.4 *Calculation of Elongation Length of Hydraulic Support Push Cylinder*

12.3.4.1 Coordinate Analysis of Chute Push–Pull Hole

According to Sect. 12.3.3.2, it can be obtained that the coordinate of the push and pull hole of the chute at section i after bending is as follows:

$$(X_i + G_2 \sin\theta_i, Y_i - G_2 \cos\theta_i)$$

Coordinate of the push and pull hole of the chute at section i before bending:

$$\left(\frac{2i-1}{2}L, \frac{a}{2} - G_2\right)$$

12.3.4.2 Analytical Model of Hydraulic Support Pressing Cylinder

Taking the analysis of the positive pull type short push bar as an example, the calculation diagram is shown in Fig. 12.8.

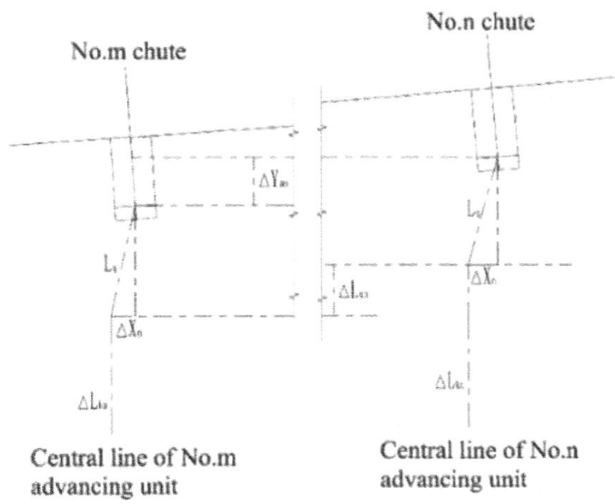

Fig. 12.8 Pressing cylinder elongation calculation diagram

Define the coordinate difference between the push–pull hole of the chute at section I after bending and the coordinate difference between the push–pull holes of the chute at section I before bending can be calculated by the following equation:

$$(\Delta X_i, \Delta Y_i) = \left(X_i + G_2 \sin\theta_i - \frac{2i-1}{2}L, Y_i + G_2(1-\cos\theta_i) - \frac{a}{2}\right) \quad (12.11)$$

The difference between the coordinate of the push–pull hole of the chute at the m_{th} section and the coordinate of the push–pull hole of the chute at the n_{th} section after bending can be calculated by the following formula:

$$(\Delta X_{mn}, \Delta Y_{mn}) = (X_n - X_m, Y_n - Y_m) \quad (12.12)$$

L_q is defined as the length of the push block of the hydraulic support, and L_{ki} is defined as the sum of the length of the push frame of the ith hydraulic support and the piston rod. Assuming that the push frame of the hydraulic support and the piston rod can only move in a straight line along the sliding trough of the hydraulic support base, then the elongation length of the pushing cylinder at the n_{th} segment relative to the mth segment can be expressed as:

$$\Delta L_{mn} = L_{kn} - L_{km} = \sqrt{L_q^2 - \Delta X_m^2} - \sqrt{L_q^2 - \Delta X_n^2} + \Delta Y_{mn} \quad (12.13)$$

12.3.4.3 Quantitative Pushing and Sliding Method

At present, most of the quantitative push sliding methods are *i/(2 N-1)* stroke of frame *i*. Assuming that the bending section is composed of 9 sections, the push sliding lengths are: 1/9 stroke of the bending section at the first section, 2/9 stroke of the bending section at the second section.

But in the actual process of pushing, is not strictly in accordance with the specifications of the passage, but also need to use this section method for calculation and correction, so as to make the scraper conveyor bending section form more reasonable.

Table 12.3 is the calculation table of the sliding length of the hydraulic support:

Table 12.3 Calculation of pushing length of hydraulic support (unit: mm)

The number *i*	Length of cylinder elongation (relative to paragraph 0)	Theoretical length of elongation	Difference
0	—	0	0
1	13.88	70	−56.12
2	53.11	140	−86.89
3	119.29	210	−90.71
4	211.62	280	−68.38
5	329.75	350	−20.25
6	441.93	420	21.93
7	529.28	490	39.28
8	591.81	560	31.81
9	629.14	630	−0.86
10	635.15	630	5.15

Note No. 0 chute is the first chute that is about to enter but has not yet entered the bending section at this time; No. 1–9 chute is the bending section chute number; Chute No. 10 is the first chute that just rolled out the bending section

12.4 Joint Positioning and Attitude Determination Method of Shearer and Scraper Conveyor

12.4.1 Coupling Analysis of Positioning and Attitude Between Shearer and Scraper Conveyor in Lateral Single Tool Operation

Because the coupling between guide slipper and pin row of shearer and the coupling between support slipper and shovel plate are carried out simultaneously in real time, and the effect of the coupling relationship between the two groups directly affects the pitch Angle of the machine, it is necessary to analyze the coupling relationship between the two groups.

12.4.1.1 Coupling Relation Between shearer's Supporting Slipper and Shovel Plate

(1) The contact form between the supporting slipper and the shovel coal plate Shearer fuselage Angle reflects the ups and downs between the front and back two supporting slipper. There are three contact modes between the left and right supporting slipper and the shovel coal plate. As shown in Fig. 12.9. The middle section is suspended. Its decision rules are shown in Table 12.4, which are defined as four modes of “0”, “10”, “11” and “2".

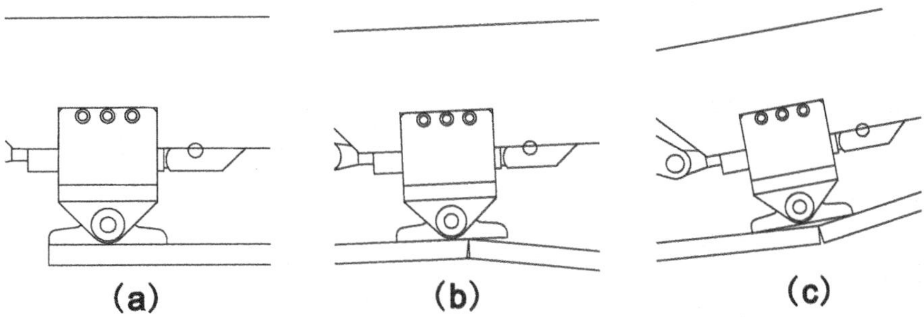

Fig. 12.9 Contact model of supporting slipper and coal plate in the middle chute. **a** Full contact, with the supporting slipper completely in contact with the middle trough. **b** Semi-contact. The supporting slipper is located at the intersection of two sections of the middle trough, and can only contact a part of the middle trough. **c** Suspended. The supporting slipper is located at the intersection of the two sections of the central trough, and cannot fully contact the two sections of the central trough

Table 12.4 Contact decision rules

Model	Meaning	Condition	Computing angle
0	Full contact, support slipper located entirely in a middle slot	(1) $Na = Nb$ (2) $Na \neq Nb$ and FloatHA[Na] = FloatHA[Nb]	Na
10	Semi-contact and in the middle slot of the interval where point A is	(1) $Na \neq Nb$ and $Na = No_1$, FloatHA[Na] > FloatHA[Nb]	Na
11	Semi-contact and in the middle slot of the interval where point B is	(1) $Na \neq Nb$ and $Nb = No_1$, FloatHA[Na] > FloatHA[Nb]	Nb
2	Dangling	(1) $Na \neq Nb$ and FloatHA[Na] < FloatHA[Nb]	Solve by using the dangling algorithm

Among them, A, B and O_1 are the left, right and middle points supporting the sole line of the sliding boot (Fig. 12.10), N_a, N_b and NO_1 are the interval

Fig. 12.10 Semi-contact analysis diagram

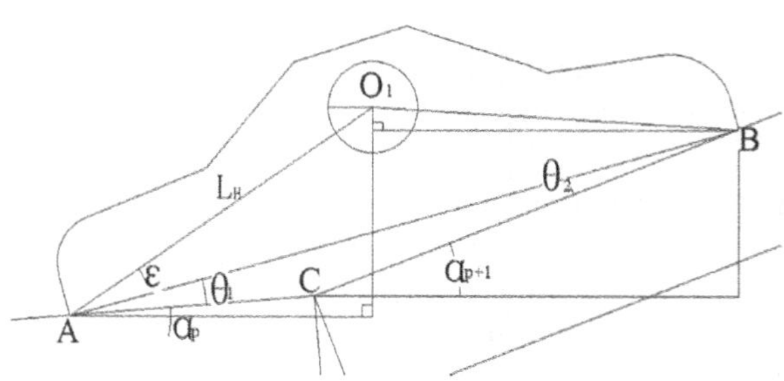

numbers of the middle trough where A, B and O1 are located, and Floatha [i] is the transverse angle of the *i*th middle trough.

(2) Analysis of contact coordinates between the supporting slipper and the shoveling coal plate.
There are three contact modes between the left and right supporting sliders and the scraper conveyor, so the combination of the three situations may have nine contact modes. Taking the most complex (c) semi-contact mode as an example for analysis, X_{O1} is the feature point of the shearer position and is a known variable, as shown in Fig. 12.10. Find the parameters at the position X_{O1}, where X_A, X_B, θ_1 and θ_2 are unknowns, and are structural parameters, where, $N_a = p, N_b = p + 1$.

According to the relationship, the following formula can be listed:

$$\begin{cases} X_B - X_A = (2L_H \cos \varepsilon) \cdot \cos(\theta_1 + \alpha_p) \\ X_{O_1} - X_A = L_H \cos(\varepsilon + \theta_1 + \alpha_p) \\ \frac{(X_B - X_C)}{\cos \alpha_{p+1} \cdot \sin \theta_1} = \frac{2L_H \cos \varepsilon}{\sin(\pi - (\alpha_{p+1} - \alpha_p))} \end{cases} \tag{12.14}$$

where: γ is the middle Angle. M_1 and M_2 are intermediate parameters set for the convenience of solving the problem. They are respectively:

$$M_1 = (-2L_H \cdot \cos(\varepsilon) \cdot \sin(\alpha_p) + L_H \cdot \sin(\varepsilon + \alpha_p) - C \cdot \cos(\alpha_{p+1}))/(X_C - X_{O1})$$
$$M_2 = (2L_H \cdot \cos(\varepsilon) \cdot \cos(\alpha_p) - L_H \cdot \sin(\varepsilon + \alpha_p))/(X_C - X_{O1})$$
$$\gamma = \arcsin(M_2 \Big/ \sqrt{M_1^2 + M_2^2})$$

Solution:

$$\theta_1 = \pi/2 - \gamma$$

$$X_A = X_{O1} - L_H \cos(\theta_1 + \alpha_p + \beta)$$

$$X_B = X_{O1} + 2L_H \cos\beta \cos(\theta_1 + \alpha_p) - L_H \cos(\theta_1 + \alpha_p + \beta)$$

In the case of (a) and (b), Y_{01} can be concluded:

$$Y_{O1} = \begin{cases} f(X_A) + L_H \sin(\theta_1 + \alpha_p + \beta) & N_{O1} = p \\ f(X_A) + L_H \sin(\theta_1 + \alpha_{p+1} + \beta) & N_{O1} = p + 1 \end{cases} \tag{12.15}$$

12.4.1.2 Coupling Relation Between Guide Slipper and Pin Bar Shape

(1) Analysis of chute pin row equation

Because the longitudinal dip angle changes very little, the connecting pin row will bend with the shape of the adjacent two middle troughs, and its pitch angle is half of the transverse dip angle of the adjacent two middle troughs.

Lateral Angle of middle pin row:

$$\theta_{Mi} = \alpha_i \tag{12.16}$$

Angle of connecting pin row transverse inclination:

$$\theta_{Ni} = (\alpha_i + \alpha_{i+1})/2 \tag{12.17}$$

Combined with the coordinates of each axle hole, the curve equation of pin arrangement can be expressed as follows:

$$\begin{cases} g_1(x) = Y_{MXP}(1) + (x - X_{MXP}(1)) \cdot \tan\alpha_1 & X_{MXP}(1) \le x \le X_{NXP}(1) \\ g_2(x) = Y_{NXP}(1) + (x - X_{NXP}(1)) \cdot \tan\frac{\alpha_1+\alpha_2}{2} & X_{NXP}(1) < x \le X_{MXP}(2) \\ \cdots & \cdots \\ g_{2n-1}(x) = Y_{MXP}(n) + (x - X_{MXP}(n)) \cdot \tan\alpha_n & X_{MXP}(n) \le x \le X_{NXP}(n) \\ g_{2n}(x) = Y_{NXP}(n) + (x - X_{NXP}(n)) \cdot \tan\frac{\alpha_n+\alpha_{n+1}}{2} & X_{NXP}(n) < x \le X_{MXP}(n+1) \end{cases} \tag{12.18}$$

Among them: $(X_{MXP}(i), Y_{MXP}(i))$ and $(X_{NXP}(i), Y_{NXP}(i))$ are respectively the coordinates of the two axle holes on the left and right side of the central trough in the first section.

(2) Coupling calculation and analysis of traveling wheel and pin row track

After the key points O_1 and O_2 of the supporting slipper are figured out, the rotation points D_1 and D_2 of the walking wheel are solved by using the transverse and longitudinal inclination angles of the fuselage. The coordinates of these two points are used to couple with the curve of the upper point of the pin row. Then use the results to verify the airframe pitch angle. If it is not suitable, adjust the longitudinal Angle of the airframe until it is satisfied.

12.4.2 Planning Software Development Based on Unity3D

The model built in UG is put into Unity3D through model repair and transformation, and virtual scene layout is carried out according to specific rules. All the algorithms above in this section are used to compile the program, and a visual human–machine

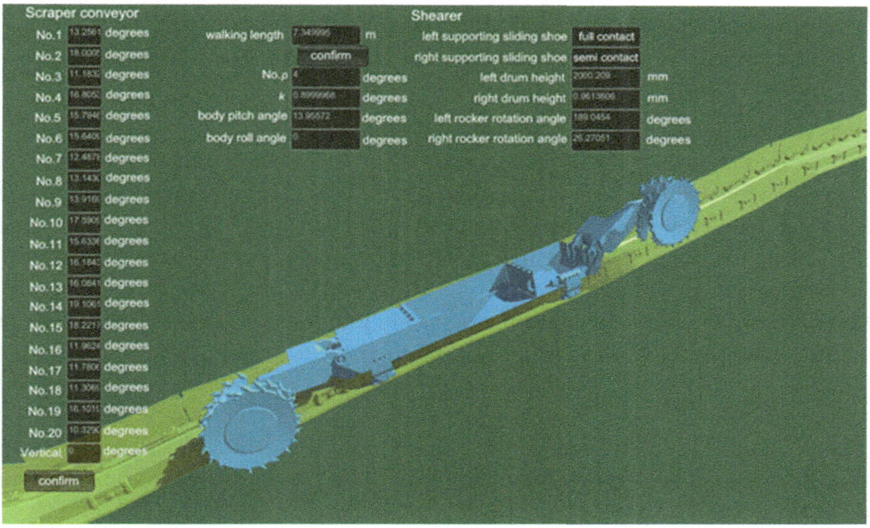

Fig. 12.11 Simulation interface of Unity 3D composite working conditions

input interactive interface is established to develop the VR planning software for the joint positioning and attitude determination of coal miner and scraper conveyor. As shown in Fig. 12.11.

In this VR planning software, the scraper conveyor inclination measured in the test process is input, and different shearer body length and structure parameters are set, then the visual simulation experiment can be carried out, and the process simulation data (shearer body pitch Angle) can be exported in XML file in real time for subsequent analysis.

Input different parameters of each section of the middle trough, calculate the shape of scraper conveyor, backstage real-time calculation of the position of the shearer walking, and in the form of coordinates to the virtual shearer, and the real-time process data stored in XML file.

The shape of the scraper conveyor can be estimated by inputting the acquired dip angles of each central trough into the virtual programming software. In order to coordinate the virtual shape of scraper conveyor, the walking position and walking attitude of the virtual shearer will be calculated in real time. Then the calculation results are transmitted to the virtual planning software screen in real time.

The running speed of the virtual shearer depends on the increment of the traction speed, and an increment is selected comprehensively in the aspects of computer calculation pressure and virtual picture fluency, so that the virtual software can carry out the planning process in real time and visually.

12.4.3 Positioning and Attitude Determination Fusion Strategy Based on Information Fusion Technology

SINS and tilt sensors are used to measure the pitch angle of the shearer fuselage and the transverse and longitudinal dip angles of each section of the middle slot. In different temperature and environmental conditions, electromagnetic interference is easy to cause sensor noise and failure. This means that the original data drift phenomenon is likely to occur on a single sensor, resulting in the actual status of the shearer and scraper conveyor marked by the sensor is not accurate. Therefore, it is necessary to use the information fusion algorithm to fuse the two measurement results of the two sensors respectively to improve the accuracy.

Multi-sensor information fusion algorithm uses multiple data collected by multiple sensors at different times to identify the actual state of two devices. The premise of adaptive fusion algorithm is batch algorithm, so it is necessary to explain it, respectively using batch estimation algorithm and adaptive weighted fusion algorithm to calculate the explanation.

The batch estimation algorithm is p measurements $[\gamma_1, \gamma_2, \ldots, \gamma_p]$, the collection is from a sensor, and the repetition interval of the same collection frequency is divided into two groups:

- When p is an odd number, the two groups divided into are $[\gamma_1, \gamma_2, \ldots, \gamma(p + 1)/2]$ and $[\gamma(p + 1)/2, \gamma(p + 1)/2 + 1, \ldots, \gamma p]$.
- When p is an even number, the two groups divided into are $[\gamma_1, \gamma_2, \ldots, \gamma p/2]$ and $[\gamma p/2 + 1, \gamma p/2 + 2, \ldots, \gamma p]$.

12.4.4 Reverse Mapping Labeling Strategy Based on a Priori Perspective

Firstly, the morphological measurement value of scraper conveyor is improved by information fusion method, and then the theoretical simulation results of the change trend of shearer pitch Angle are obtained by inputting VR programming software. It is regarded as the prior knowledge, and the information fusion value of the two sensors is used to mark the prior knowledge and correspond to each other in the operation of the shearer. In particular, the position of some key inflection points should be judged and corrected from time to time, which the reverse mapping is marking strategy based on a priori perspective.

First, according to the analysis results of the two sensors, the position of the shearer is inferred. The specific method is as follows:

- According to the trend of the curve, the theoretical curve is divided into several blocks and several stages, and the key points are marked;
- When the actual value and the theoretical value change trend is the same, judge the shearer to enter a new stage, the key points are corrected;

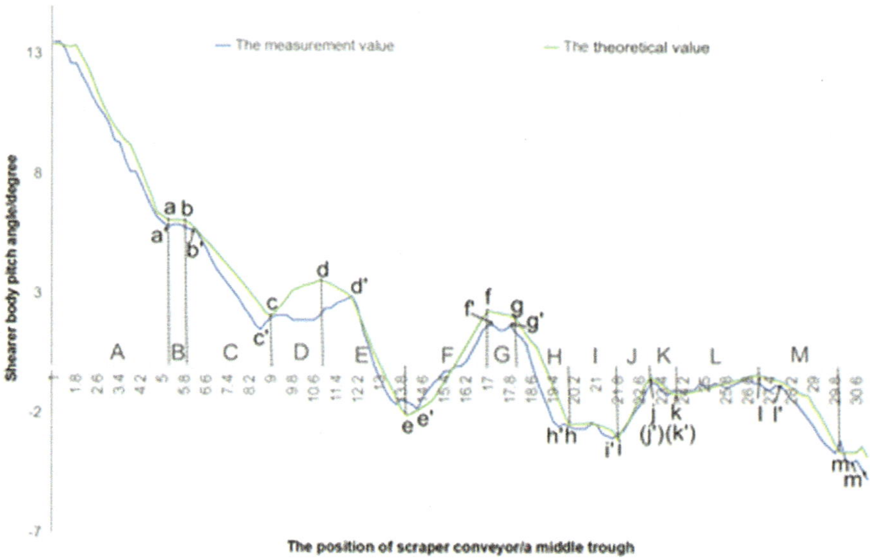

Fig. 12.12 Tagging the policy

- The actual value and the theoretical value at the stage of the key points continue to correspond and correct, and then the actual value of the reverse mapping to the scraper conveyor;
- From the nose to the tail of the machine, complete analysis and corresponding, so as to reverse push out the position of the shearer in real time.

As shown in Fig. 12.12, this case is used as an example to illustrate the reverse mapping markup strategy.

First, the theoretical curve is divided into A, B, …, M interval, and find the interval demarcation points a, b, …, m, as a priori experience. In the actual operation of the shearer, using the a′, b′, …, m′ and other points to real-time correction and verification of a, b, …, m and other points, determine the corresponding interval, and then reverse mapping, so as to find the corresponding point, so as to determine the position of the shearer on the scraper conveyor.

12.5 Method of Memorizing Posture Between Groups of Hydraulic Supports

12.5.1 Thought Source of Memory Posture of Hydraulic Support

12.5.1.1 Sources of Ideas and Methods

In the whole process of hydraulic support, the cutting track of the upper drum in the first few cycles of the shearer determines the supporting state of the roof of the back several cycles of hydraulic support, while the cutting track of the lower drum determines the forward state of the support. Therefore, under the principle of "memory cutting" theory, the group hydraulic support can also be judged and operated according to the action of similar law.

But with "Memory cutting method", is the outstanding difference between interval is equal to the passage of a cycle of fully mechanized working face coal winning machine cutting deep, so the cycle of shearer is relatively easy to determine, and length of a hydraulic support top beam support for five to seven commonly cut deep length, the base has three or four deep cutting length, its operation cycle is different, changing law is more complex. The cutting height of the previous several cycles of the shearer is used together to act on the hydraulic support, so the hydraulic support should be more "insensitive" to the change of the terrain than the shearer.

So, by similar memories of shearer cutting theory method to predict the overall group of hydraulic support posture, to be able to predict in advance to the next cycle of various hydraulic support posture adjusting range, by comparing actual attitude and predict the attitude deviation, to show the running status to normal or not, so that the accidents can make more timely and effective judgment. This section calls it "group hydraulic support memory attitude method".

12.5.1.2 Horizontal and Vertical Forecasts

The memory attitude method of hydraulic support includes both the prediction in the transverse cycle and the prediction in the longitudinal cycle. Among them, the memory attitude in the transverse cycle: it means that the support attitude which has been moved can be used to predict the support attitude which has not been moved or is about to be moved within a cycle, reflecting the change of the transverse terrain of the whole working face. Memory posture in the longitudinal cycle: it means that within several cycles, the overall state of all the supports in the next cycle can be predicted through the overall state of all the supports in the previous several cycles, reflecting the topographic changes of the entire working face along the advancing direction of the working face. The principle is shown in Fig. 12.13.

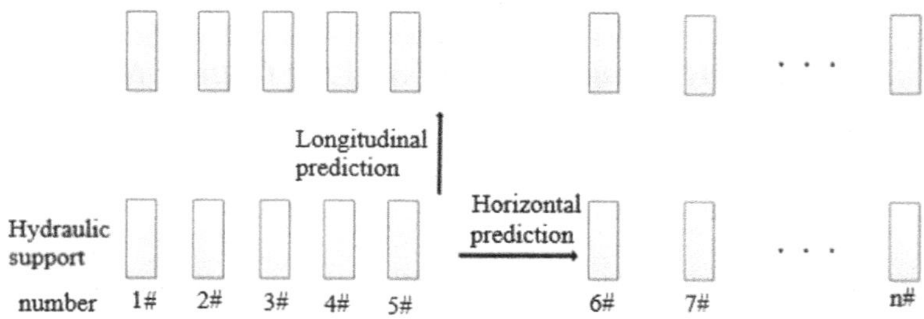

Fig. 12.13 Horizontal and longitudinal prediction of memory attitude of hydraulic support

The measured attitude data are compared with the predicted data in real time. If the difference is large, the corresponding hydraulic support is likely to be in a relatively abnormal state. Real-time measurement data is used to roll the prediction and revise the theoretical model, so as to improve the prediction accuracy more accurately.

12.5.1.3 Memory Height and Memory Angle

The memorizing attitude method of group hydraulic support is divided into the memorizing attitude of the support height and the memorizing attitude of the key Angle of inclination.

(1) Memory height
Using the cutting height of the shearer in several adjacent cycles, the height of the hydraulic support was solved, and the height value of the next cycle was predicted by using the height value of each solved cycle sequence.
(2) Memory perspective
Use (1) to solve the memory height, and then use formula inversion to solve each key Angle.

The pitch Angle of the top beam of the actual underground hydraulic support should not exceed plus or minus 7° degrees. The changes of the pitch Angle of the top beam and the dip Angle of the base are more sensitive than the changes of the terrain, so they are not suitable for the prediction of the memory attitude, but the changes are very small. Therefore, on the premise that the height can be predicted, the inclination Angle of the front and rear connecting rods of the hydraulic support can be predicted. It should be noted that the inclination Angle of the rear connecting rod and the pitch Angle of the top beam described herein are the relative angles relative to the base inclination Angle.

12.5.2 Analysis of the Relationship Between the Support Height of Hydraulic Support and the Cutting Roof Track of Shearer

12.5.2.1 Determination of Pitch Angle of Top Beam

The number 1–12 represents the cutting order. For ZZ4000/18/38 hydraulic support, the top beam and the front beam have crossed a total of 6 cutting depths from 7 to 12, and the pitch Angle of the top beam is determined by the mining height corresponding to the six cutting depths.

To judge the size of the six cyclic mining heights, find the two minimum height value serial numbers, which are denoted as M (the smallest) and N (the second smallest from the last). H(m) and H(n) represent two mining heights respectively.

- $H(m) < H(n)$, it can be judged that the hydraulic support is in the upturning state;
- $H(m) > H(n)$, it can be judged that the hydraulic support is in the down-bent state;
- In the special case, $H(m) = H(n)$, and the top beam is in a horizontal state.

Figure 12.14 shows the Angle formation process of the top beam of the hydraulic support. The pitch Angle of the top beam can be calculated by the following formula:

$$\alpha_{ij}^{zj} = \arcsin \frac{H(m) - H(n)}{J \cdot (m - n)} \tag{12.19}$$

where, J is the cutting depth.

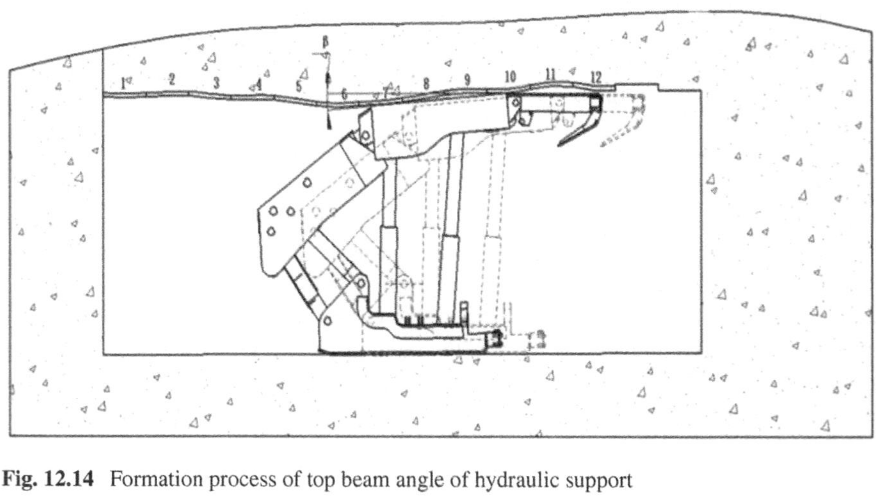

Fig. 12.14 Formation process of top beam angle of hydraulic support

The support height of the hydraulic support can be obtained by the following formula:

$$H_{H_{ij}}^{zj'} = H(m) - (J \cdot (m - 5) \cdot \sin(\alpha_{ij}^{zj})) \tag{12.20}$$

12.5.2.2 Height Solution and Antithesis Program

After solving the angle of the top beam, the angle of the front connecting rod and the angle of the base can be obtained. The hydraulic support can be placed according to the solution height to determine the placement method of the whole virtual scene.

The hydraulic support height formula can be expressed as:

$$H_{zj} = f(\alpha_{ij}^{zj}, \beta_{ij}^{zj}, \gamma_{ij}^{zj}) \tag{12.21}$$

In this section, δ_{ij}^{zj} is assumed to be 0°, therefore, α_{ij}^{zj} can be solved by the following equations:

$$\begin{cases} f(\alpha_{ij}^{zj}, \beta_{ij}^{zj}) = H(m) - (J * (m - 5) * \sin(\alpha_{ij}^{zj})) \\ \alpha_{ij}^{zj} = \arcsin \frac{H(m)-H(n)}{J*(m-n)} \end{cases} \tag{12.22}$$

12.5.3 *VR Monitoring Method of Memory Posture*

VR scene can display the running state of the whole group of hydraulic supports in the form of 3D in real time. Mainly rely on real-time data sent back from the working face. This requires the establishment of virtual models and virtual scenes that are completely consistent with the actual scaffolds. The algorithm and formula are programmed into the program, and the interface is reserved to read the real-time working face data, so as to keep the real-time synchronization with the working face state.

12.6 Summary

This chapter of "Compressors" of fully mechanized working face in actual working condition of the constraints of the connection between relations are studied, and the attitude behavior of the composite study, including the ideal floor flat cases "compressors" virtual collaborative technology, coal winning machine and the coupling between the scraper conveyer under different environment-coal winning machine and scraper conveyor of the posture under bending section of the feed coupling, coal winning machine and scraper conveyor under complicated working condition of joint positioning of informant method and the mutual influence between group of hydraulic support posture-group of hydraulic support memory monitoring method.

Chapter 13
Virtual Monitoring System

13.1 Introduction

After the "Single machine" condition monitoring and virtual simulation method and the "Three machine" condition monitoring and simulation method in VR environment, on this basis, we need to design the monitoring system of "Three machine" working condition in the whole VR environment. The application of VR technology in fully mechanized mining monitoring has more intuitive and reliable VR monitoring advantages and potential.

This system comes from the "Digital Twin" theory in "industry 4.0". Based on the research of the theory, it integrates with the actual monitoring, and designs the whole system under the Unity3D simulation engine.

13.2 Digital Twin Theory of Fully Mechanized Working Face Equipment

13.2.1 Introduction of Digital Twin Theory

The digital twin model refers to the digital presentation of physical objects in virtual space, that is, to create a virtual model for physical objects in digital way, to make full use of the data such as physical model, sensor update, operation history and other data., to integrate the simulation process of multidisciplinary, multiple physical quantities, multi-scale and multi-probability, to complete the mapping in virtual space, and to simulate its behavior characteristics in the real environment, and then reflects the whole life cycle process of the corresponding entity equipment.

X. Wang et al., *Virtual Reality Technology in Mining Machinery*,
https://doi.org/10.1007/978-981-16-4408-5_13

Generally speaking, it refers to copying a physical object in a digital way, simulating the behavior of the object in the real environment, realizing the virtualization and digitization of the whole process, so as to solve the past problems or achieve accurate forecast of the future.

Siemens applies it to the design of unmanned factory, which can achieve the following functions:

- All links in the production process are simulated and analyzed;
- In the stage of design and selection, we can see the whole production process;
- Plan the operation details and strategies to improve efficiency;
- Predict the possible problems and optimize the whole system;
- Try to test everything at the beginning;
- Integration of model selection design and process planning.

In the process of realizing digital twin, three necessary conditions are needed:

(1) On the premise of not affecting the normal work, corresponding measures are taken on the physical entity, appropriate sensors are arranged, and the characteristic variables are collected. After special processing and algorithm, the status of the equipment can be accurately obtained;
(2) In the virtual world, the virtual image of an object needs the ability to simulate the state of the corresponding real physical entity in the virtual world;
(3) "Real" and "Virtual" interface: how to accurately and reliably transmit the state variables collected in (1) to (2), and can be seamlessly received by (2), and make corresponding actions according to this information. Keeping data synchronization with real physical entities is called VR monitoring.

13.2.2 Digital Twin Fully Mechanized Face Equipment

Establish "Three machine" digital information model, including "Three machine" digital model, "Three machine" information model and "Virtual" and "Real" interface, as shown in Fig. 13.1.

Among them, the "Three machine" information model is a digital prototype based on the design theory and method of the "Three machine" itself and the digital expression of real physical products. The attitude of the "Three machine" under the actual working condition is analyzed, and the synthesis method of synchronous characteristic variables of the "Three machine" state is obtained. The information fusion algorithm is studied, and the "Three machine" reliability monitoring theory and method are obtained law.

The virtual digital model of three machines in fully mechanized mining face is mainly a virtual image which is completely consistent with the actual state of three machines in fully mechanized mining face. In the virtual environment, each component and its surrounding components are constrained and defined, and the interface variables are reserved, which can receive the running state data sent back by three machines in fully mechanized mining face in physical space in real time.

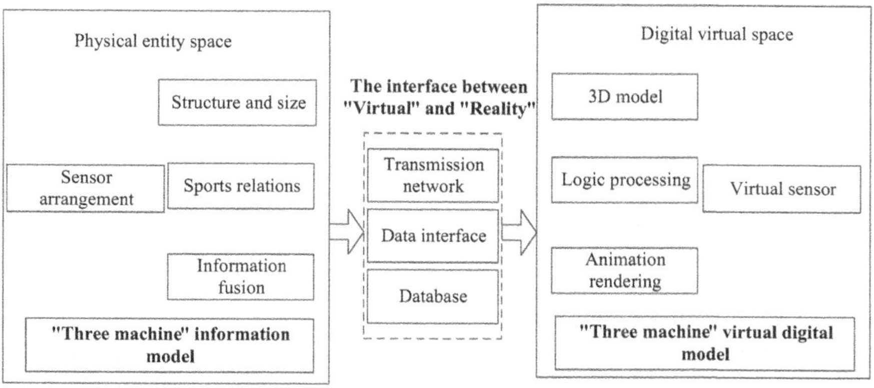

Fig. 13.1 A "Three machine" digital twin theory system

The interface of "Virtual" and "Reality" can combine the information model and virtual digital model. Through the high-speed network communication platform, it can receive the data sent back by the sensors arranged on the three machines in fully mechanized mining face in real time, transmit it to the database, and transmit it to the central control center of crossheading, remote dispatching room and the information model interface of VR monitoring host in VR laboratory. This interface mainly corresponds to the physical information sensing system in Chap. 2 and the main content of VR monitoring system in this chapter.

In this way, the operation status of the three machines in fully mechanized mining face can be accurately simulated and real-time monitored and synchronized. The operator can interact with the system, enter any area of the system simulation at any time through any space, observe the operation condition of the equipment, give an alarm to the abnormal situation, and timely find and deal with the faults and problems existing in the operation.

13.3 Overall Framework Design of VR + LAN "Three Machine" Condition Monitoring System

13.3.1 System Design Objectives

This monitoring system is based on VR technology, through the data sent back by the sensors of the equipment in the fully mechanized working face in real time, the corresponding VR monitoring picture is driven in real time, and the virtual image of the "Three machines" in the fully mechanized working face in real time is established, so that the real-time running state of the working face can be displayed in a three-dimensional panoramic view with low delay, clear picture and no jamming.

According to the demand analysis of the system, the main design objectives of the system are as follows:

- The VR model and environment which are completely consistent with the real Three machines in fully mechanized mining face are established, and the corresponding control scripts are written, so that the actual Three machines in fully mechanized mining face running state can be fully displayed in the virtual screen;
- The virtual "Three machine" model needs to reserve the interface of characteristic variables of each sensor, and it can easily import the real-time working face equipment data which has been stored in the database;
- Sensors that can reflect the position and posture change of Three machines in fully mechanized mining face shall be arranged on the actual working face, and can be reasonably arranged on all equipment without affecting the normal operation of the equipment;
- In the monitoring process, the data should be stored and the historical data can be fully analyzed and modeled from the perspective of big data, so as to predict the operation status of the whole fully mechanized mining equipment;
- In the monitoring mode, because of the large number of equipment in the comprehensive mining face, the amount of sensing information is large. If only one monitoring host is owned, the server pressure will be huge, the VR monitoring screen will not run smoothly or there will be a carton site, which seriously affects the reliability of VR monitoring. Therefore, it is urgent to study a technical solution to solve this problem.

13.3.2 Hardware Design

On the basis of the early intelligent transformation of the experimental system of the complete set of fully mechanized mining equipment in our laboratory, some sensors are installed on the shearer, hydraulic support and scraper conveyor respectively, and a high-speed network communication platform of wireless sensor network and wired network is established to transmit all kinds of sensor signals to the server (including PLC signal, electro-hydraulic control system signal and combined switch and so on.).The system hardware network structure design is shown in Fig. 13.2.

The shearer body sensor is connected to the mine intrinsically safe wireless base station through the onboard PLC and reserved RJ45 interface, which is sent back to the centralized control center.

13.3.3 Software Design

The software system is designed by using the large VR simulation engine Unity3D, which has a variety of human–computer interaction hardware device interface, network cooperation, reserved interface with database and other modules, and is

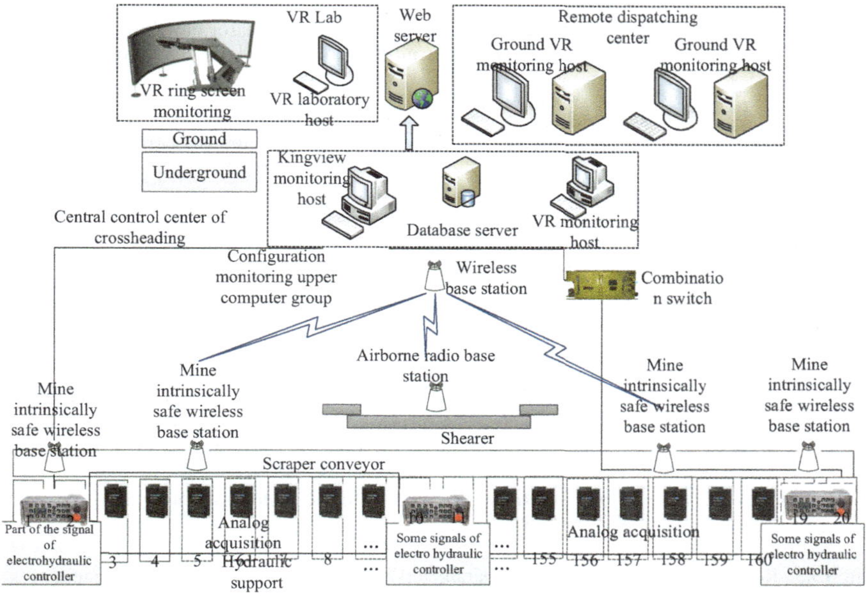

Fig. 13.2 System network structure diagram

programmed and designed by VS2013 and C # language. The data collaboration module uses SQL server as the database for real-time data storage, XML file as the auxiliary record module, and MATLAB as the operation module.

Kingview6.55 is used as the upper computer for data acquisition in the software system. When the network I/O module is connected to the mine intrinsically safe base station KT113-F, it can be connected to the high-speed network communication platform which has been arranged to completely cover the fully mechanized mining face, so that the signal can be collected and sent back to the central control center of crossheading, and then it can be connected to the Kingview monitoring host through Modbus TCP protocol, and the Kingview monitoring system can upload the collected data to SQL in real time Server2008 database, VR monitoring host can call the data in the same LAN database in real time to complete the connection of data to VR host.

13.3.4 Real Time Sensing System

The real-time equipment operation parameters collected by the real-time sensing system of the VR monitoring system in this book are shown in Fig. 13.3. In addition to the attitude parameters collected by the sensors arranged in the physical information

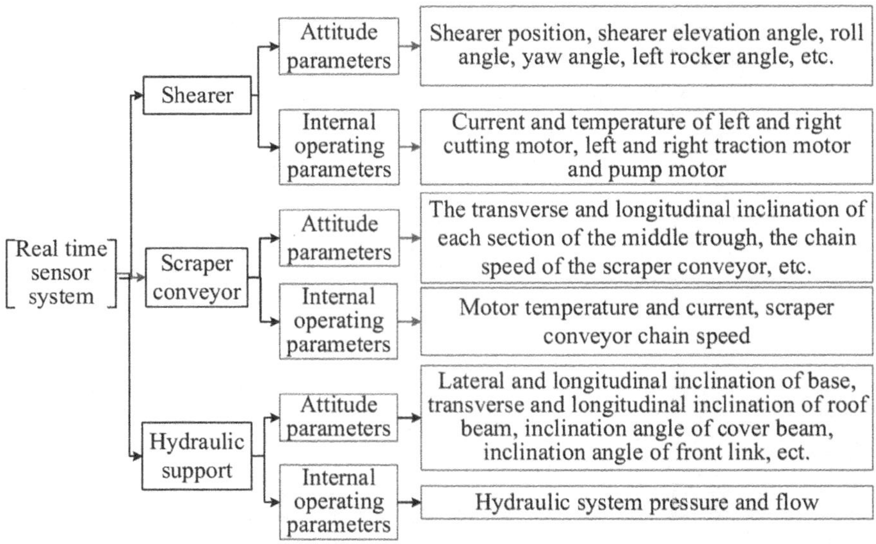

Fig. 13.3 Real time equipment operation parameters collected by real time sensing system

sensing system, the internal operation parameters (such as motor current, temperature and other information of each equipment) can also be collected.

13.4 VR Monitoring Method Based on Unity3D

VR monitoring scene can display the operation status of the whole fully mechanized working face in 3D form in real time. It mainly depends on the data sent back by the working face in real time. Therefore, it is necessary to establish the virtual model and virtual scene which are completely consistent with the actual fully mechanized mining scene. Only when the algorithm and formula are programmed and the interface is reserved to read the real-time data of working face, can it keep synchronization with the equipment state of working face.

13.4.1 Reservation of State Variables in VR Environment

In order to control the movement and state of virtual devices, virtual variables need to be reserved. By controlling virtual variables, virtual devices can be controlled to operate. The overall steps are as follows: make the virtual parts correspond to the defined virtual variables, then compile the attitude analysis results, conversion angle and fusion algorithm into the C# program, real-time sensor data is assigned to the virtual variables, and then manipulate the virtual parts to move, and the virtual equipment will present the real motion state in real time.

In the aspect of model motion, all the motion is completed by real-time calculus, and the animation method is not used to ensure the restoration ability of the actual scene to the maximum extent. After the model is built, its movement is controlled by the script attached to the parent–child relationship. For the part that needs precise control, a motion command with parameters can be written in "FixedUpdate()" method to realize precise control of motion by controlling parameters.

Through the reserved interface, we can achieve the following two purposes: one is to operate the virtual device by changing these variables; the other is to monitor the running status of the device by displaying the value of variables in real time.

13.4.2 Real Time Data Reading and Access Method

Establish various signal tables in SQL Server table to receive the data transmitted back by Kingview in real time. The VR monitoring program has a database interface, which can connect with the database in real time and read the data. The data is transmitted to the virtual variables through the virtual agent. So as to drive the corresponding virtual device to carry out the corresponding action.

13.4.3 Real Time Calculation Method of Underlying Mathematical Model

In order to make the monitoring more reliable and full of reality, we must make full use of the "Three machine" real-time data obtained, and carry out the calculation in the background and bottom layer, mainly using the methods in Chap. 3 to compile into DLL file in Matlab program and link the bottom model. Then the prediction results are displayed on the screen and compared with the actual running results. The types of underlying models are shown in Fig. 13.4.

13.4.4 Real Time Rendering of Mining Environment Information

Underground environment modeling mainly includes intelligent real-time rendering of roof and floor curves. According to the real-time shearer attitude information and cutting conditions, the cutting height of the front and rear drum is calculated in real time, and then the cutting roof and floor information is obtained, and updated in

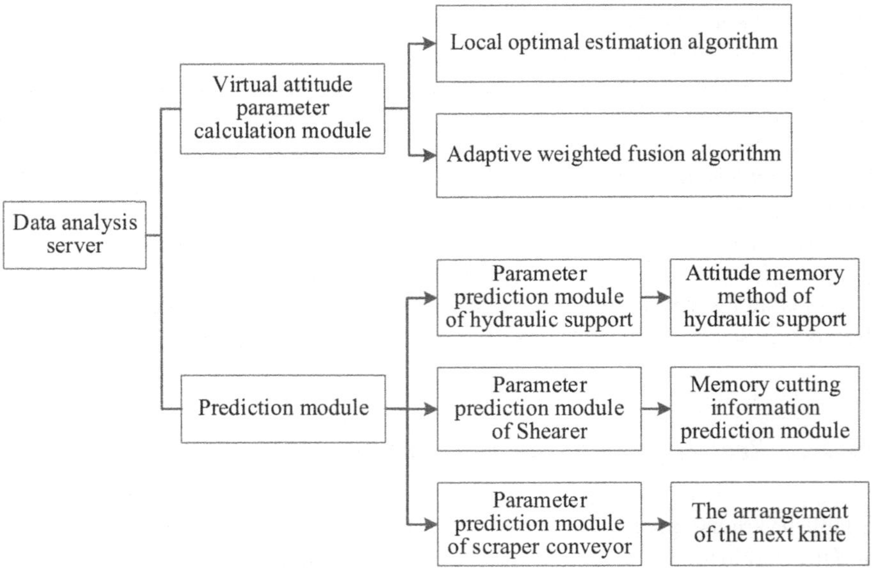

Fig. 13.4 Prediction module using historical data for analysis

XML file in real time. The virtual environment reads the data in the XML file and renders it to get the virtual roof and floor curve.

Each cutting top and bottom curve is connected by each cutting key point, and each cutting key point is drawn with a small circle point. The specific method is to define the virtual object array Dingban of cutting key points, add the objects to the scene list, combine with XML data, generate them in real time, and connect them with line command. The connecting lines of every two adjacent cutting points are smoothed by using NURBS or Beizer curves.

13.4.5 Fault Occurrence Screen Representation

According to the real-time data of the working face, judge the current operating conditions and possible monitoring problems. After the current problems are determined by the monitoring system, render the corresponding problem effect in VR monitoring scene in real time, vividly and timely show the current problems to the underground operators and remote dispatching room personnel, and give the prompt opinions of remote manual intervention, which can provide reference for VR monitoring of underground fully mechanized working face The monitoring screen with early warning function is provided.

13.4.6 Implementation of Real-Time Switching Video Monitoring Screen in VR Environment

Some specific recommended virtual camera positions will appear in the virtual picture of VR monitoring program. Click any virtual camera position, and the operation condition of the whole working face will be displayed from this perspective. When the device is dangerous or has problems, the system can quickly lock the virtual camera near the problem device to watch, and quickly find and solve the problem.

The location of the virtual camera is mainly based on the location of the video monitoring layout in the existing more advanced underground automatic intelligent mining equipment. For example: the head and tail of scraper conveyor, the cameras arranged every certain number in the hydraulic support, and the cameras arranged on the shearer body to observe the cutting situation of the front and rear drum.

13.5 Virtual Monitoring and Real Time Synchronization Method Based on LAN

The cooperation mode of each monitoring host in the VR monitoring upper computer group is shown in Fig. 13.5. They respectively obtain the information from the corresponding data server, receive the VR monitoring data of other corresponding devices for real-time synchronization, and share their own data to the network view.

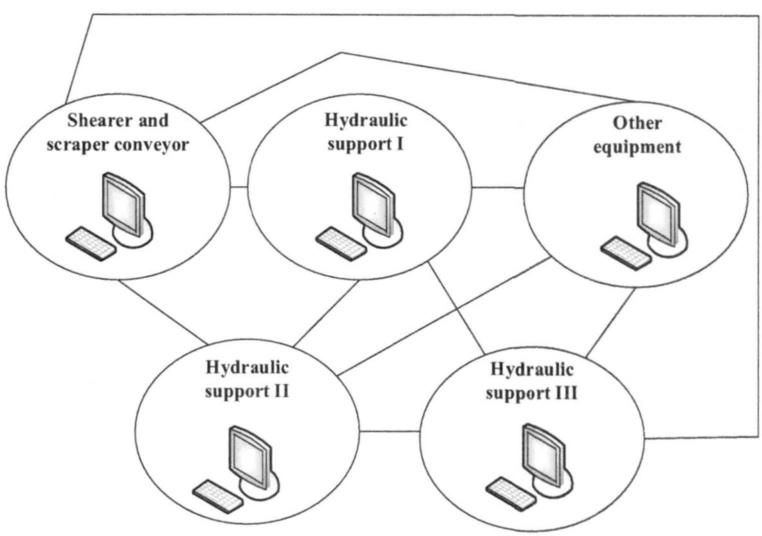

Fig. 13.5 Cooperation mode of VR monitoring hosts

Taking the VR monitoring program of shearer and scraper conveyor as an example, the virtual LAN collaborative interface is to send the real-time data of shearer and scraper conveyor to the VR monitoring program of other equipment through RPC command in "NetworkView" component of Unity3D, and receive the data of corresponding equipment from VR monitoring program of other equipment to drive the project The virtual models of several other equipment in the working face can act accordingly to synthesize a panoramic picture of the whole working face, so that the monitoring host installed with VR monitoring program of shearer and scraper conveyor can monitor the whole panoramic fully mechanized working face. The working principle and information exchange diagram are shown in Fig. 13.6.

The specific scheme of LAN cooperation method is as follows: take VR monitoring program of shearer and scraper conveyor as server, VR monitoring procedure of hydraulic support I, VR monitoring procedure of hydraulic support II, VR monitoring procedure III of hydraulic support and VR monitoring procedure of other equipment as client connection, and the steps are as follows:

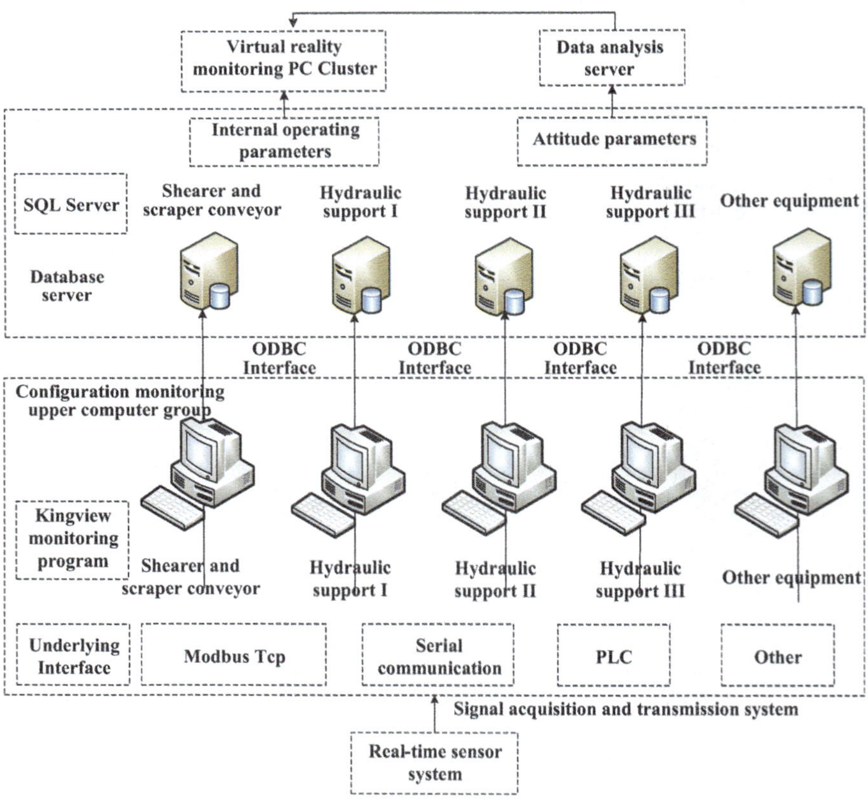

Fig. 13.6 Interaction diagram of VR monitoring host of shearer and scraper conveyor and other relevant information

Firstly, the server is established in VR monitoring program of shearer and scraper conveyor

NetworkConnectionError error = Network.InitializeServer (30, port useNET);

NetworkConnectionError error = Network.InitializeServer (30, port useNET);

The VR monitoring procedure of hydraulic support I, VR monitoring procedure II of hydraulic support, VR monitoring procedure III of hydraulic support and VR monitoring procedure of other equipment are connected to the server through the following instructions:

NetworkConnectionError error = Network.Connect(ip, port);

So all five parts of VR monitoring program are connected to a LAN;

VR monitoring program of shearer and scraper conveyor will obtain control right of virtual shearer and virtual scraper conveyor in virtual scene, and associate and read virtual variable with variables in cmjgbj table in SQLSERVER database server. The variables and meanings of shearer database are shown in Table 13.1, through the following code:

cmd.CommandText = "SELECT * FROM CmjGbj where ID = (select MAX(ID) from CmjGbj)"; //Get the latest data from the database
if (reader.Read()){CmjPoistion = reader["CmjPoistion"].ToString();//Read the latest location data of shearer;

The control right of different objects is obtained according to a certain allocation mode, which is mainly marked by the value of controltag variable, as shown in Table 13.2:

VR monitoring program of shearer and scraper conveyor sends virtual shearer running state variables to VR monitoring program of hydraulic support I, VR monitoring program of hydraulic support II, VR monitoring program of hydraulic support III and VR monitoring program of other equipment in real time through RPC command of networkview.

Table 13.1 Variables and meaning of shearer database

Serial number	Variable name and key words in database	Meaning
1	CmjPoistion	Shearer position data
2	CmjFuYangJiao	Pitch angle data
3	CmjHengGunJiao	Roll angle data
4	CmjPianHangJiao	Yaw angle data
5	CmjYouGunTongZhuanJiao	Left drum angle data
6	CmjYouGunTongZhuanJiao	Right roller angle data
7	CmjZuoYouGangShenChang	Left high cylinder extension data
8	CmjYouYouGangShenChang	Right high cylinder extension data

Table 13.2 Serial number and control right acquisition of ControlTag

Serial number	ControlTag numerical value	Acquired control
1	ControlTag=0	Virtual shearer and virtual scraper conveyor
2	ControlTag=1	The first 50 virtual hydraulic supports
3	ControlTag=2	The 51st–100th virtual hydraulic support
4	ControlTag=3	Virtual hydraulic support for rest
5	ControlTag=4	Other virtual devices

```
if(ControlTag==0){CmjGbjTransform_basic = GameObject.Find("CmjGbj").GetComponent<CmjGbj>().CmjGbj_Transform;
CmjGbjTransform_basic.GetComponent<NetworkView>().RPC("ReceiveControl",
RPCMode.AllBuffered,CmjPoistion,CmjFuYangJiao,CmjHengGunJiao,
CmjPianHangJiao,CmjZuoGunTongZhuanJiao,CmjYouGunTongZhuanJiao,
CmjZuoYouGangShenChang,CmjYouYouGangShenChang);}
```

In the hydraulic support VR monitoring program I, the hydraulic support VR monitoring program II, the hydraulic support VR Monitoring Program III and other equipment VR monitoring program ends, the data in the shearer and scraper conveyor VR monitoring program are obtained through the following commands. Through the following code implementation:

[RPC].

```
void ReceiveControl(float CmjPoistion,CmjFuYangJiao,float CmjHengGunJiao,float CmjPianHangJiao,float CmjZuoGunTongZhuanJiao,float CmjYouGunTongZhuanJiao,float CmjZuoYouGangShenChang,float CmjYouYouGangShenChang){
CmjPoistion = CmjPoistion_Others;//Shearer position synchronization
CmjFuYangJiao = CmjFuYangJiao_Others;//Shearer pitch angle synchronization
......//Other data synchronization of shearer}
```

Synchronous control and display is completed by setting the pose attribute of virtual Shearer to ref type, that is, when calling and changing data, the original file data is changed directly;

```
void OnSerializeNetworkView(BitStream stream, NetworkMessageInfo info)
{if (stream.isWriting) {float CmjPoistion = CmjPoistion_Others;stream.Serialize(ref CmjPoistion);}
Else{ float CmjPoistion = 0;stream.Serialize(ref CmjPoistion);//Shearer position data update.
......//Other data update of shearer}
```

13.5.1 Collaboration and Data Flow Based on RPC Technology

In order to realize the network cooperation function, the whole scene must be networked, as shown in Fig. 13.7. Specifically, each virtual device is networked. From the interface variables of each device established above, the virtual collaboration of Three machines in fully mechanized mining face is actually the collaboration among the virtual variables of each virtual device. Each object has a management node, which is responsible for managing each device. Each device corresponds to a virtual image in the corresponding networkview. The virtual images of each device can interact with each other, and each virtual image corresponds to the corresponding management node, so as to feed back the data to the device.

All devices add "NetworkView" components to complete data synchronization through RPC function. This function can be called locally or remotely. It is very flexible and easy to call. The function is executed directly at the called place. Because synchronization of network view has returned program health at all times, asynchronous call is adopted here, that is, after the call, it does not wait for the execution result to return.

In order to ensure that the networked system is as close to the actual system as possible, the system adopts the authorized server mode, all operations are concentrated on the server running program. The client has the power to obtain information and input commands, and does not directly participate in the operation, so that the

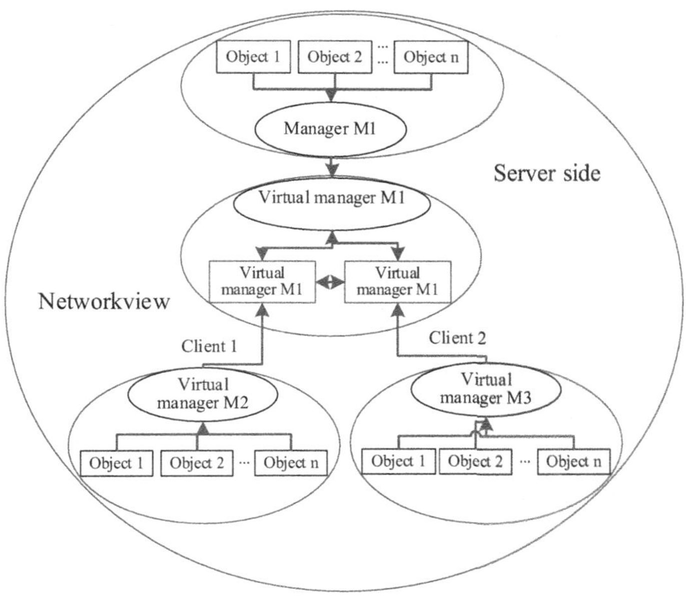

Fig. 13.7 Network collaboration framework

system can be guaranteed to run stably to the maximum extent. All the objects' motion parameters are controlled by an RPC function and can be called remotely at any time.

13.6 Real Time Coupling Strategy of Multi Software

Network I/O module connects all sensing information to high-speed network communication platform, and accesses to Kingview monitoring host through Modbus TCP protocol. The Kingview monitoring system can upload the collected data to SQL Server 2008 database in real time. VR monitoring host can call data in a local area network in real time, and then complete the connection between data and VR host. As shown in Fig. 13.8. Then, Unity3D, MATLAB and SQL Server interact in real time.

13.6.1 Kingview + SQL Server

Each part of Kingview monitoring program collects the corresponding device data from the signal acquisition and transmission system in real time, and transmits it to the corresponding database through ODBC interface.

Take the configuration monitoring program of shearer and scraper conveyor as an example for analysis. First, after the ODBC interface is established, write the connection code in Kingview software, open the monitoring interface, and set the insertion frequency, the data can be transmitted to the database server in real time.

The deviceid is the "DeviceID" generated when connecting to the database, which remains unchanged during the connection. When storing the data, according to the unique device identification between the upper system and the server, create the form templates "ceshiCmj" and "ceshiGbj" in the configuration system, establish

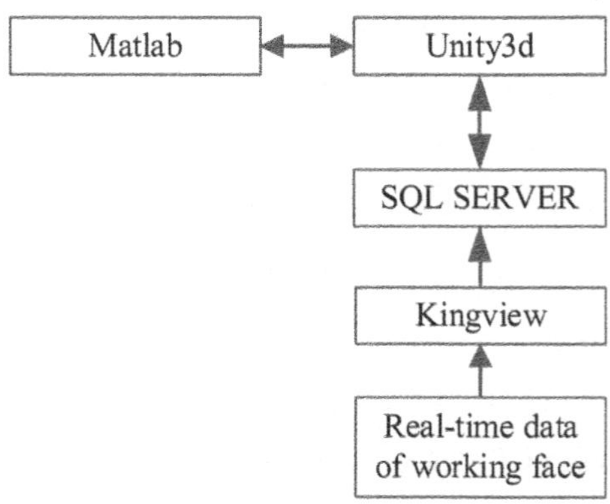

Fig. 13.8 Software interaction diagram

the record body "Cmj" and "Gbj", and then associate them through the "SQLInsert" command, and then associate the related tables and record body with the "SQLConnect" table in the SQL server through the "SQLConnect" command. "SQLCommit" means to insert the data into the data In the library. In this chapter, the insertion frequency is set to 200 ms.

13.6.2 SQL SERVER + Unity3D

Unity3D uses "Start ()" and "Update ()" functions to program. Among them, "Update ()" is called every time a new frame is rendered, and the call speed is inconsistent according to the computer configuration and picture quality;" FixedUpdate ()" is executed at a fixed time interval, not affected by the frame rate; therefore, this chapter selects "FixedUpdate ()" for event update.

After the data can be transferred into SQL Server database in real time, Unity3D software needs to interact with the data in real time by using the interface written in C #. During the operation of VR monitoring program, the corresponding reading and updating frequency can be set. For example, if the updating frequency is 200 ms, the meaning is to call the data five times in 1 s. After reading the information, it is processed according to the information fusion algorithm, and then transferred to the virtual model for movement. As shown in Fig. 13.9.

13.6.3 Matlab Calculation Result Processing

The data analysis server is a high-performance server integrated with MATLAB software and Unity3D software, and can obtain the data of the database server in real time for analysis, including virtual attitude parameter calculation module and prediction module, and each module has compiled the algorithm and published it as dll. file; the virtual attitude parameter calculation module is based on multi-sensor information fusion technology, Using multiple data of a sensor in a period of time, using a specific algorithm for calculation and multiple sensor data with correlation for secondary information fusion, the accuracy of attitude parameter data can be improved to the maximum.

13.7 Prototype System Development

VR monitoring can be compared with the real video monitoring system of the underground comprehensive mining face, which can supplement the real-time monitoring of the comprehensive mining operation state.

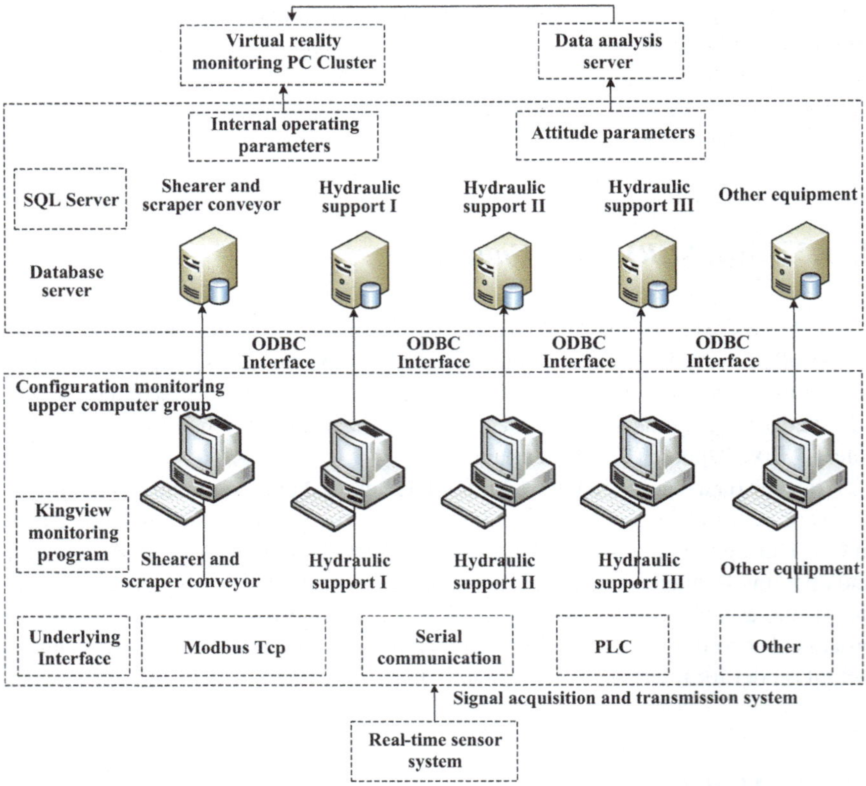

Fig. 13.9 Working diagram of configuration monitoring upper computer group and database server

- The VR monitoring system of three machines in fully mechanized mining face in VR + LAN environment proposed in this book is essentially a VR monitoring method and system based on LAN cooperation. Compared with the existing technology, the following functions can be completed:
- This VR monitoring method monitors the operating conditions of fully mechanized mining equipment in LAN environment, adopts modular design, and the virtual equipment reads the secondary high-precision attitude data of the corresponding monitoring equipment stored in the database server in real time through various algorithms and drives; and the cooperation between the modules is through the underlying protocol in the LAN, which effectively solves the problem The pressure of the server is large, and the VR monitoring screen is not running smoothly or stuck;
- In the process of VR monitoring, the configuration software is used to store data in the database, and the data analysis module is set up. The historical data is used to analyze and model, and the operation status of fully mechanized mining equipment is predicted;

- The VR monitoring method integrates data into the monitoring, establishes a high-quality integrated panoramic workbench, and comprehensively and systematically monitors the equipment operation of the comprehensive mining face in real time.

13.8 Summary

This chapter introduces the "three machine" condition monitoring system of fully mechanized working face under VR environment. Firstly, the integration of "digital twin" thought and fully mechanized working face equipment is analyzed, and then design the whole system, including system design objectives, software design, hardware design and function design. Then, the six key technologies of the VR monitoring method based on Unity3D, such as variable reservation, virtual interface, and the real-time synchronization method based on LAN are studied, then, the real-time coupling strategy of "Kingview + SQL Server + NetworkView + MATLAB" is studied, provide software technical support for VR monitoring, finally, the prototype system is designed and introduced.

Chapter 14
Summary and Conclusions

14.1 Work Summary

(1) With "virtual reality+" coal machine equipment, "Internet+" coal machine equipment and so on as the main line of research, the deep "synergy—fusion—transformation—innovation" research of coal machine equipment and virtual reality was carried out. To screen system as the hardware support, integrated dual channel column with Visual Studio, Open Scene Graph and UG software as the software platform system, integrated with force feedback device, data glove and human–computer interaction equipment such as position tracker from virtual simulation element model conversion technology and the establishment of the model library, the method of virtual assembly based on OSG, method of virtual assembly based on UG, method of virtual assembly human–computer interaction, virtual assembly five aspects of network method oriented to virtual reality machine equipment assembly aspects were studied.

(2) The method of single machine condition monitoring and virtual simulation for fully mechanized mining face is studied. In the part of working condition monitoring, the actual problems of flat floor and horizontal and longitudinal inclination Angle in the operation of equipment are studied. The physical information sensing system is established to seek for accurate and reliable working condition monitoring method of single machine. In the aspect of virtual simulation, virtual single machine which is completely consistent with the actual three machines in fully mechanized mining face is simulated under the environment of VR simulation engine Unity3D, including attitude analysis, model building and repair technology. For each single machine operation process of the existence of special cases, including hydraulic support components seamless linkage, scraper conveyor virtual bending, virtual memory cutting of shearer and other methods are specially studied.

(3) The method of three machines in fully mechanized mining face condition monitoring and virtual simulation for fully mechanized mining face is studied. "Compressors" were analyzed under the condition of the underground actual

X. Wang et al., *Virtual Reality Technology in Mining Machinery*,
https://doi.org/10.1007/978-981-16-4408-5_14

working condition between the constraints of the connection relations, study the complex relationship between the attitude behavior, and with the help of the real-time sensor signal and the establishment good "compressors" VR software, respectively for the ideal of bottom case "compressors" virtual collaborative research, coal winning machine and scraper conveyor bending section of the feed attitude coupled behavior research; The joint positioning and attitude determination method of shearer and scraper conveyor under complex working conditions and the memory attitude monitoring method of group hydraulic support are also studied.

(4) In view of the situation that the Three machines in fully mechanized mining face sensor nodes have many signals, which causes great pressure to the transmission network and the monitoring host, the overall intelligent monitoring scheme under the "VR + LAN" environment is proposed, and the Three machines in fully mechanized mining face working condition monitoring system under the VR environment is established. Of "Digital Twin" thought and equipment integration of fully mechanized working face is analyzed, then carries on the overall design of the system, including the system design goal, design of hardware and software design and function design, etc., to VR monitoring method based on Unity3D variables in reserve, the key technologies of virtual interface and so on six big and virtual monitoring and real-time synchronization method based on LAN, and then to "kingview + SQL SERVER + Net Work View + Matlab" software more real-time coupling strategy is studied.

(5) In view of the static and single parameters of the existing automatic control program of Three machines in fully mechanized mining face, which cannot well meet and match the complex and changeable dynamic environment, a collaborative mathematical model of "three machines" in fully-mechanized mining under dynamic environment is established, and a VR collaborative planning method based on MAS (FMUNITYSIM) is proposed. The multi-factor coupling three machines in fully mechanized mining face collaborative mathematical model, three machines in fully mechanized mining face Agent model, and three machines in fully mechanized mining face VR planning method are established. In view of the changes and differences of the working environment, equipment, process and other dynamic characteristics, the three machines in fully mechanized mining face co-operating VR system is designed by multi-factor from multi-dimension. The key parameters of the three machines in fully mechanized mining face are planned and adjusted online, and the collaborative virtual simulation under different parameters is realized, and the design of the best parameter matching method is realized, which provides a theoretical basis for the rapid planning and safe production of the fully mechanized mining face.

14.2 Main Conclusions

(1) Table 14.1 shows the comparison results of the system established by the three virtual assembly methods in Chap. 4 based on OSG, Chap. 5 based on UG, and Chap. 7 based on virtual assembly network in terms of hardware requirements, software development requirements, cross-platform, and the degree of refinement of model.

According to the comparison in Table 14.1, it can be seen that the system established by the three methods has its own characteristics. It should be selected according to the needs of different users and the requirements of the three systems.

- Users who have high requirements for 3D display effect of virtual reality and human–computer interaction can choose virtual assembly method based on OSG;
- It requires a high degree of refinement and other mechanical properties of the model, so it is recommended to choose the virtual assembly method based on UG.
- According to the requirements of establishing virtual assembly system in network environment such as LAN or Internet, it is suggested to choose virtual assembly network method.

Table 14.1 Comparison results

Compare the project	Virtual assembly technology and system of coal machine equipment based on OSG	Virtual assembly technology and system of coal machine equipment based on UG	Network technology and system of virtual reality assembly for coal machine equipment
Hardware requirements	Higher	High	The lower
Software development requirements	Higher	Higher	Higher
Cross-platform nature	Poor	Bad	Excellence
The refinement of the model	Good	Pretty good	Bad
Mechanical properties	Use the coordinate position constraint	You can add	Use the coordinate position constraint
The human–computer interaction(HCI)	strong better	poor	poor
3D display effect	Pretty good	Good	Medium
Network synergy	Medium	Poor	Good
Programming Requirements	High	High	High
The degree of practical	Suitable for stand-alone environment	Suitable for stand-alone environment	Applicable to the network environment
Portability	Poor	Bad	Good

(2) On the monitoring methods of three machines in fully mechanized mining face

(a) In this paper, a method of solving the joint posture of the coal mining machine and scraper conveyor is proposed in this paper. In this paper, the relationship between the two is analyzed, the analysis of the coupling relationship, the virtual simulation, the multi-sensor information fusion and the marking strategy are used to study the real-time morphological coupling relationship between the top Angle of the coal mining machine and the conveyor. By means of the experiment, the error of the positioning of the coal mining machine is not within the length of the slot of 0.38, and there is no cumulative error, which can be more detailed and accurate dynamic monitoring for the operating condition of the coal mining machine and scraper conveyor.

(b) In view of the problem that the bending section of scraper conveyor cannot be solved accurately, a method of calculating the attitude of the bending section chute with the analytical method of key dimension coordinates is put forward. By using this method, the expansion length of the bending section chute, the pin row and the pushing cylinder and the change relationship between the yaw Angle of the coal miner and the walking track can be solved precisely. The correctness of the method is verified by experiments, which can guarantee the running state and stable reliability of the shearer in inclined cutting, and also provide a theoretical basis for the collaborative correlation and three-dimensional positioning of the "three machines" in fully mechanized mining face.

(c) To solve the problems of lack of overall monitoring method for group hydraulic support attitude and lack of immersion in 2D monitoring, a memory attitude method for group hydraulic support in VR environment was proposed. The internal relationship between the mining height and the hydraulic support height is solved, and the real-time 3D visualization monitoring is carried out.

(3) About VR scene simulation of three machines in fully mechanized mining face

(a) In view of the lack of virtual simulation methods such as spatial positioning of virtual miner, bending of scraper conveyor and four-link linkage of hydraulic support, a virtual co-simulation method of horizontal ideal bottom plate of three machines in fully mechanized mining face was proposed. The motion states, tasks and relationships based on the finite state machine theory are established. The virtual collaboration/perception method of Three machines in fully mechanized mining face, the virtual walking method of coal miner, the virtual bending method of scraper conveyor and the seamless linkage method of hydraulic support parts were studied, and two interactive methods of GUI interface and virtual hand operation were established, which improved the technical level of Three machines in fully mechanized mining face virtual simulation.

(b) In view of the problems that geological and topographic changes and uncontrollable simulation process model are not fully considered in the current three machines in fully mechanized mining face virtual simulation, a virtual shearer memory cutting method is proposed. In Unity3D environment generates the virtual environment of roof and floor, and then the virtual scraper conveyor in virtual floor laid as virtual coal winning chance the track, click on the virtual operating button for virtual coal winning machine, the virtual controller real-time operating data storage, analysis, processing and read again, realization of coal winning machine cutting simulation of memory. This method can quickly test the new memory cutting technology algorithm under different geological environment, and can realize the training of shearer driver alone and the cooperative operation and training of two people.

(4) VR monitoring of fully mechanized mining face

(a) in view of the lack of a mature "compressors" collaborative monitoring model, this paper puts forward a kind of Digital virtual monitoring model of fully mechanized working face of Twin theory, including the information model, the virtual Digital model and interface of virtual environment and the actual environment, realize the fully mechanized equipment in the whole life cycle management and link the height of the Digital and modular.

(b) Aiming at a series of problems in VR monitoring system, such as lack of mathematical model of support system, less utilization and mining of historical data, and high pressure on server caused by large amount of sensor information, a VR monitoring method based on LAN collaboration was proposed. The mirror image of the real fully mechanized mining face was established under the environment of Unity3D, and the real-time data collected by the monitoring host was stored into the SQL Server2008 database. VR equipment was driven in real time through the attitude calculation module and prediction module embedded in the background. Through C/S architecture and LAN collaboration, interactive real-time synchronization is realized among multiple hosts. Finally, a panoramic picture of the whole working face is synthesized, providing a reliable distributed computing solution for fully-mechanized mining panoramic VR monitoring.

(c) On the fully-mechanized mining face three machines in fully mechanized mining face virtual planning

Aiming at the problems of single parameters of three-machine virtual simulation, no mathematical model as support, data cannot be exported and exchanged, and the poor response between virtual simulation parameters and real parameters, a MAS based VR collaborative planning method was proposed. "Compressors" of fully mechanized working face under the dynamic environment is established mathematical model of collaborative research of shearer, scraper conveyor, hydraulic support group, for the underground hydraulic system and dynamic environment, and other methods of communication, coordination and conflict resolution between the Agent and redundant processing problems such as perception, and established the collaborative planning system, based on online on these compressors “key parameters for planning and control”. By means of the co-simulation under different parameters, the motion supporting relationship of the “three engines” can be vividly displayed, and the planning situation of the “three engines” can be viewed from multiple angles in real time. The design of different test schemes, statistics and analysis of the three machines in fully mechanized mining face under different conditions of the key parameters of operation, the best matching parameters, provides a theoretical basis for the rapid planning of fully mechanized mining face and safety production.

The manufacturer's authorised representative in the EU is Springer Nature Customer Service Centre GmbH, Europaplatz 3, 69115 Heidelberg, Germany. If you have any concerns regarding our products, please contact ProductSafety@springernature.com

Printed and bound by CPI Group (UK) Ltd, Croydon, CR0 4YY
30/09/2026
02225295-0006